ZAGREB ONE FOUR

ZAGREB ONE FOUR

CLEARED TO COLLIDE?

Richard Weston and Ronald Hurst

Jason Aronson/NEW YORK

For Ruth Weinreich Pedersen
Stewardess, British Airways

For the one hundred and seventy-five
persons who died with her at Vrbovec, Yugoslavia
on 10 September 1976

and for
Gradimir Tasic
Air Traffic Controller
who also became a victim

Foreword

by

Sir Peter Masefield, M.A., D.Sc., D.Tech., Hon.F.R.Ae.S., F.C.I.T.

On the 10th of September, 1976, at 10.14 hours and 38 seconds precisely, over Vrbovec in Yugoslavia, there happened one of the most tragic air disasters in recent history.

A British Airways Trident 3 – G-AWZT – with 54 passengers and a crew of nine bound from Heathrow to Istanbul collided at 33 000 feet with an Inex-Adria DC-9 – YU-AJR – with 108 passengers and a crew of seven, flying from Split to Cologne.

This million-to-one chance, in which 176 people were killed, came about from an error on the part of the Zagreb Air Traffic Control Centre. Subsequently the controller concerned – Gradimir Tasic – was tried, found guilty and sentenced to seven years in prison.

Through a chance encounter and a budding friendship, a British lawyer, Richard C. W. Weston, an experienced private pilot and a Member of the Flight Safety Foundation went to Zagreb to represent the family of Ruth Pedersen, one of the British Airways stewardesses killed in the accident.

During the course of the trial of the air traffic controllers concerned with the accident, Richard Weston realised that more than just those responsible were on trial – instead the focus was not only upon the ATC system but also upon human responsibilities, professional accountability and the system of justice as well. The International Federation of Air Traffic Controllers entered the lists and, at the centre of it all, Richard Weston took upon himself the unexpected mantle of a champion of justice and of fair play for both the relatives of the victims of the accident and for Gradimir Tasic, and his other air traffic control colleagues in the dock.

The result of a saga, unique in aeronautical annals, was that Tasic was released from prison in November, 1978, after he had served 27 months of his sentence.

The authors have written a remarkable account of the events which led up to the accident and what happened after it – a book which is at once macabre and compelling, carrying, as it does, a record of human endeavour and compassion, of technical misadventure and the involvement from Britain of judicial processes in a distant Balkan state.

This book has had no parallel and, in its central theme, we may hope will never have a sequel. I commend it for both its human content and for its extraordinary background. Richard Weston became involved in a remarkable trail of events: Ronald Hurst had determined that the record should be made public.

Between them they have produced a book which is a classic in its field.

P.G.M.

Authors' Notes

It will be helpful to bring certain matters to the attention of readers.

This book is about an accident. The material on which this account is based includes the report produced by the Aircraft Accident Investigation Commission of the Yugoslav Federal Civil Aviation Administration[1] – but not the official British reprint of that report[2] since permission to quote from this was not granted – contemporary news coverage: tape and documentary records of the accident: and the records of the enquiry and of the trial which followed.

In certain cases, therefore, there are different versions of the same event: for example, the tape recording of conversation by pilots and/or controllers may give a time A, which may subsequently be reported as time B, referred to in the enquiry as time C and at the trial, as time D.

Many of these discrepancies are important links in the chain of events: rather than pepper every page with footnotes, however, we have chosen to ask readers to exercise their own vigilance and to evaluate these differences accordingly.

The authors are aware that there is no such language as Serbo Croat or, as it is occasionally described in the text, Serbo-Croatian. These terms in fact indicate two different languages: nevertheless, they occur in many of the source documents, are widely used and for that reason are allowed to stand. In these documents, too, we have encountered a number of variations in the transliteration of names from the Cyrillic to the Roman alphabet: the choice here has been made in an effort to achieve simplicity.

Certain of the points raised during the trial of Gradimir Tasic are reported verbatim. These questions and statements appeared in the International Press. There was, however, no specific transcript of the evidence but only a summary dictated by the Judge at appropriate intervals. In addition certain translations were personally initiated by Richard Weston.

The Court-room dialogue in this work essentially interprets the information which is contained in this material and in the Judge's Findings.

In air traffic control a strip of paper is used with the essential details of the flight

[1]Komisija za Ispitivanje Avionskih Nesreća Savezne Uprave za Civilno Zrakoplovstvo.
[2]British Airways Trident G-AWZT: Inex Adria DC9 YU-AJR. Aircraft Accident Report 5/77 HMSO London.

written thereon. This strip of paper measures approximately 8 inches long by 1 inch wide.

In the United Kingdom Manual of Air Traffic Services and in general, this piece of paper is referred to as a strip. However, in this book it is also called a slip: the terms are synonymous and have been used as they appear in their context, whether printed as verbatim copy or recorded as the spoken word.

In reporting numbers to each other, controllers and pilots invariably use the full phonetic pronunciation of each digit; thus what may be colloquially spoken to describe Finair Flight 1673, would between controller and pilot be stated thus 'Finair One Six Seven Three'. Similarly 22 minutes past 10 would be reported in aviation language as One Zero Two Two. A flight level would be referred to as Two Five Zero, Two Eight Zero, Three One Zero or Three Five Zero as the case may be. When there is a great deal of this, as for example in Chapters Four and Six, it not only clutters up the page but makes the reading rather laborious, particularly when the sequence is an historical sequence without any actual emphasis being placed on the specific numbers. The authors have therefore, in the large majority of cases used the ordinary numeric symbols. The actual phonetic sounds of each digit have only been used where particular emphasis is to be placed on them. The professional aviator or controller will no doubt when reading deal with these in whatever way he wishes. The importance to the lay reader however, is to remember that it takes longer to say these things when they are spelled out, than it does when the numeric form is used. So 22 minutes past 10 takes less time to say than One Zero Two Two, and this is important in the sense that the transmission times which are printed out by the computer-driven tape are the times when a message *starts*. It is impossible to state the exact duration of any particular radio/telephone or telephone message in the text, but if a reader wishes to make some calculations the question of enunciation set out above is, of course, important.

In aviation matters the common standard for time around the world is Greenwich Mean Time. People landing in New York at 3 o'clock in the morning will find that the time has been logged as 0800 hrs. because that is the GMT standard. It is only important in this case in the sense that all the tape recordings quoted have referred to the period of time shortly after 10 o'clock, which was Greenwich Mean Time; but colloquially, and in the Trial, the local time of 11 o'clock was used. Once again, the authors have deliberately retained the original figures which were used and the reader is asked to accept that in those circumstances 10 and 11 were in fact one and the same time.

Acknowledgements

It would be impractical to attempt to list the many friends, colleagues and associates who assisted so generously during the preparation of this book. Some contributions, however, were made under exceptional circumstances; they are noted here, therefore, as special debts.

First, to the many people in Air Traffic Control who so gladly offered help and advice. Their aid was enlisted through the good offices of Tom Harrison, then General Secretary of the International Federation of Air Traffic Controllers' Associations who also extended an especially warm welcome to this project on behalf of the President and Executive Board.

It should be said in this regard that the authors have nevertheless beaten their own path. The views expressed herein are those of Richard Weston and Ronald Hurst and do not necessarily represent the policies of IFATCA or, indeed, of any other body. It is hoped, however, that in essence this endeavour will earn the support of those concerned for the safety of international aviation: not least among these is the ordinary member of the public.

The description of K.P. Dom Zabela in the Prologue is based on the editorial in *The Controller* 1/79.

Dragomir Modrusan, Richard Weston's lawyer in Zagreb, is the 'Drasko' of this book. Both he and his sister, Zeljka afforded continuous assistance and hospitality over the long course of the trial and the subsequent legal processes. Their help with translations and advice on the Yugoslav legal system was invaluable.

Much is owed, too, to Kay Coombs and the late Rosemary Hayer, both of Her Britannic Majesty's Consulate in Zagreb for their assistance during the period from 10th September, 1976 and the dark months which followed: similarly, thanks are due to Captain Stuart Clarke who undertook a considerable journey in the quest for information: to Alison Pilpel for a timely intervention and for equally painstaking research; to David Liddiment, for permission to quote from his own report on the Zagreb collision . . . and to John Cutler, ARAES, who gave his time and skill to the production of the diagrams in this work.

Cossor Electronics provided the photograph of the radar screen: Margaret Crisfield typed the manuscript . . . and David Fulton, of Granada Publishing waited with quite an extraordinary forbearance.

The authors are indebted to Peter Wilde of London Air Traffic Control Centre, West Drayton, for his help in reading the script.

All of these kindnesses are recorded with appreciation.

Prologue

For those who do not know of this place, K.P. Dom Zabela is located near Povarevac, Yugoslavia, some 200 kilometres from Belgrade.

K.P. Dom Zabela is a prison.

Yet it is only the two-metre high barbed wire fence enclosing the site which betokens its true function, for within the area, the buildings are anonymous and office-like, muting the brief impressions of the visitor who counts his stay in minutes.

Nor is the atmosphere in the waiting room vastly different from that to be found in any workers' cafe in any European industrial city. It is only the ugly grey-flannel dress and military-style cap worn by those who queue at the counter for soft drinks, the dirty shirts under this unappealing uniform and the unshaven faces which focus the mind on the common plight.

Under protest or in resignation, those who are here by decree have exchanged one life for another, forfeiting home and family and friends for the drab reality which now surrounds and submerges them.

It is, of course, no more than the price commonly demanded by the society they have outraged: at worst, for punitive discipline and a bare cell, and at best, for the shared misery of a small room, the round of petty, unwanted and resented tasks and the unending anguish of the captive spirit. It is affliction: but as yet we do not know of another way to salve the wounds these shuffling men have inflicted on us.

Among these men is the air traffic controller, Gradimir Tasic, for in his case, too, the law has followed its course, found him guilty after due process and made him, therefore, as one with these others; save that their offences are perhaps better defined and the logic of their punishment more readily comprehended.

There can be no such reconciliation for this man who is judged both innocent and guilty: innocent – for all concerned with his trial are agreed on the absence of criminal intent – yet guilty, by virtue of negligence in the exercise of his duties, of contributing to the death of one hundred and seventy-six human beings.

It is for this last bitter reason that the law has made its disposition of Gradimir Tasic. It is for that reason that, on this twenty-fourth day of November in the year 1978, he is numbered as a prisoner at K.P. Dom Zabela.

His sentence, it is true, is claimed not to be an act of revenge. It is, rather, the strict and necessary application of the law, following a fair and public trial before international observers in a civilised country; and, if the order and integrity of that trail is beyond dispute, what then is this man's claim on the public conscience? What need to challenge the verdict of the courts, and what need, given his proven share of responsibility for the strewn wreckage of two airliners and that frightful toll of lives, to question the wisdom, nature and meaning of the judgement which has brought him here?

It is ironic that the need is implicit in the sad comment of Yugoslavia's Deputy Minister of Justice, Gojko Prodanic, that 'the State has never considered Tasic to be a criminal', for, if this is true – and it is tragically so – then he is not, therefore, punished for any of the acts of mischief or malevolence which may mark out his fellow prisoners. He is punished for having been placed under intolerable pressure, for being unable to bring order to a technological nightmare, for being no more than human in those circumstances and for being defeated by that humanity.

An obscure exile, this K.P. Dom Zabela, far from the daily joys and adventures of those who are free. Yet Tasic's presence remains vivid in the mind of every one of the world's air traffic controllers sitting at his console, his image too easily reflected in those glowing screens which chart the passage of the hosts of aircraft and the trusting millions seated within them. It is an uncomfortable and restless image: a wan ghost which has no place in the considered deliberation necessary to this trade; but while that presence is there, that calm lies too thinly over a disturbing thought, clouds the controller's judgement and so reaches into the lives of every person aboard a civil airliner.

What happened to Gradimir Tasic can happen to me.

Chapter One

On 10 September 1976 a Trident aircraft of British Airways, flying in clear skies at a height of 33 000 feet above the Zagreb radio beacon, collided with a DC9 of the Yugoslav charter airline Inex-Adria. All aboard both aircraft - 176 people - were killed.

It is a fact of the tragic record that within hours of the crash, eight members of the Zagreb regional flight control centre were arrested and charged with criminal negligence: that seven of them were subsequently acquitted and that one, air traffic controller Gradimir Tasic, was found guilty and sentenced to seven years' imprisonment.

It is generally accepted that the purpose of judicial retribution is to protect society and in this case, following the most exhaustive trial and the proven guilt of the accused, the apportioning of punishment would appear to be no more than a sad and necessary postscript to one of aviation's most terrible disasters. Yet it is Tasic's ordeal which brings this book into being for, far from protecting society, the effect of that decision handed down by the District Court of Zagreb bears heavily - and adversely - on the safety of every air traveller who must depend on the fidelity and efficiency of the international air traffic system.

It is the authors' purpose to show that these qualities have been seriously compromised by that baleful verdict and that the implacable destruction of one man's professional career is no corrective for the inherent technological deficiencies which combined to bring about his downfall. The brutal facts are that the deficiencies of the system remain and that, as a consequence, aviation has acquired a new and unnecessary hazard: simply, that in any moment of potential crisis, the air traffic controller's essential decisions may no longer be governed solely by his professional evaluation of the situation. They must be influenced, too, however fleetingly, by his knowledge of the fate of Gradimir Tasic; and by the possibility that the next and irrevocable act or omission might cause him to share that fate.

For the air traveller, for the public at large and for all those concerned with the safety of international civil aviation, the importance of that momentary hesitation may yet be made cruelly plain.

The origins and aftermath of the Zagreb disaster are therefore examined in these pages in order that the implications of risk may be widely understood by all those who - however unwittingly - must now share them; for only this

15

knowledge can ignite the determination to secure improvements necessary to the betterment of safety in this field.

In the area of air traffic control, this understanding on the part of the public has hitherto been clouded by the comparative obscurity in which the controller carries out his duties: an obscurity compounded both by the arcane nature of his task and by the manner in which, in the minds of the public, the whole concept of aviation is popularly embodied in the formidable figure of the airline pilot.

This concept has been nurtured since the beginning of commercial aviation, for it is on the pilot and his personal odyssey that attention has been focussed; and it is his image with its corona of exciting and noble endeavour which is most readily associated with its progress. For these reasons, the triumphs and the tragedies of the man at the controls have been recorded in the utmost detail, and his special professional skill acknowledged, both in his privileged uniform and in the respect which society, conscious of the responsibilities and qualities thereby represented, duly accords its wearer.

There is no similar tradition of epic or even minor literature to distinguish the equally vital role of the air traffic controller, or to illustrate the gravity of his responsibility for the safe conduct of ever greater numbers of aircraft through the crowded air lanes. Nor is there a wide public knowledge or even a unanimity of understanding among his employers of the absolute nature of the decisions he must make, or any comprehension of the conditions of urgency and physiological and psychological stress under which he must so often make them. And least of all, there is no outward symbolism of gold braid and insignia, either to reward him with the deference automatically extended to his pilot colleague, or to testify that on any working day, hundreds, and often thousands, of human lives will depend on his professional competence. Quite literally thousands; for some measure of that dependence may be gleaned from the fact that peak traffic at a major airport can mean that aircraft generally carrying any number up to 400 passengers are landing and taking off at a rate in excess of one per minute; while the controllers responsible for an airspace such as the United Kingdom's London Terminal area, for example, may invigilate the transit of some two and a half thousand aircraft during an equally busy period of 24 hours.

In this work the reader is asked to consider the situation and behaviour of a number of air traffic controllers during the critical minutes of a disaster in the making. For this reason it is necessary to offer some basic description of the frame in which the episode was enacted, and some indication of the human and environmental factors which imposed, and continue to impose, their own influence on the pattern.

The ATC System

Air Traffic Control is a ground-based service dedicated to the achievement of the safe, orderly and expeditious movement of air traffic. Its essential aim is to prevent collision, either in the air, or during manoeuvres such as taxiing or towing on the ground, by providing instructions, guidance and advice to pilots through the medium of radio telephone (R/T) voice communications.[1] In the majority of cases ATC personnel are trained and employed by the National Civil Aviation Administration although in a number of countries the service is integrated with the military system or entirely operated or supervised by the military authorities.[2]

Reference has already been made to the great volume of traffic at a major airport. The complexity of the control function becomes even more apparent when it is realised that in addition to the demands of airliners in transit, and of the airline and airport operators to whom they are of immediate concern, the service – and the word should be regarded as a euphemism for the individual controller – must recognise and be ready to meet the requirements of military, executive and private aviation. The controller, further, must adapt his expertise to aircraft cruising speeds ranging from that of a light executive machine flying at 150 miles per hour to that of the wide-bodied jet transport flying at speeds up to and sometimes exceeding 600 m.p.h., having performance characteristics in the climb and descent phase often in excess of 5 000 feet per minute and the capacity to close with other aircraft at speeds exceeding 1 000 miles per hour.

The problem of ensuring safety and expedition in the face of mathematics of this order is therefore a three-dimensional one of time, geographical position and height compounded by continuing movement. The answer lies in the use of procedural control: the application of separation standards to fit these three dimensions. Thus, aircraft are separated from each other vertically, using height, longitudinally, using time and laterally, using geographical separation. These separation standards are internationally agreed and in effect may be seen as the protective boundaries sheltering each aircraft: the art of controlling, of course, is to ensure that these boundaries are not infringed – that is, that they are not encroached upon by other aircraft.

In order to do this, therefore, a vertical separation of 1 000 feet is called for in the case of aircraft flying below 29 000 feet. Above that height due to the possibility of altimeter error at greater altitudes, a separation of 2 000 vertical feet is accorded. Aircraft flying down each airway in opposite directions must also maintain pre-set flight levels appropriate to their heading. Thus on the airway in which the crash occurred the British Trident flying eastwards was flying at 33 000 feet, but could also have been given flight levels of 27 000, 29 000 or 37 000 feet. The DC9, flying in the opposite direction, would have been given

[1]For this purpose English is designated by I.C.A.O. (the International Civil Aviation Organisation) as the official language of aviation and is used in all international ATC operations. In recent years a disquieting element has been introduced, however, whereby nationalist pressures in the Province of Quebec have succeeded in securing a Government recommendation legalising bilingual (French – English) air traffic control services in that region. Pilots and anglophone air traffic controllers have expressed opposition and concern at the threat to safety standards. See also footnote, page 22.
[2]Argentina, Uruguay, Italy, Brazil (see IFATCA circular December 1977).

the alternate flight levels – 26000, 28000, 31000 and 35000 feet.

A flight plan must be prepared for any flight, and each sector through which the aircraft is going to fly is given an information slip in advance which sets out the significant details of the flight. These 'flight progress slips', based on telexed and telephoned advice and on computer print-out, are arranged on the procedural controller's console; as each aircraft enters his sector and reports its position and height, he uses the information on each slip to co-ordinate and direct other aircraft in the sector to make sure that each is correctly and safely positioned both for the present and the immediate, or foreseeable future. Radar supplements the use of the slips by identifying each plane using the 'Squawk' system described below.

The importance of the flight progress slip to the air traffic controller should be particularly noted here: the effect of an alleged failure to provide such a slip is a point to be considered in this work.

In relative terms the basic separation standards are wasteful of airspace in busy airspace complexes. For example, the standard longitudinal separation of two aircraft at the same speed on the same track is ten minutes flying time. This related to the present generation of passenger-carrying aircraft can represent a distance exceeding 50 miles. However, to assist the controller in his task and to reduce the distance between aircraft, thus permitting greater utilisation of the airspace, the technology of radar control has been developed to a high degree of efficiency, whereby displayed on the controller's radar screen is not only a response – the 'blip' or target – denoting the presence of an aircraft, but also, alongside that response (as shown in Plate 1), the alphanumeric symbols signifying its identity or call-sign and, in many cases, the height at which it is flying. (It is of special relevance here that this type of *complete* identification is provided from the aircraft in reply to the controller's instruction that it should 'Squawk' – i.e. transmit – a specific code number.[3]) As a result of the use of radar it is possible to provide a different form of separation enabling the distance between aircraft to be reduced to approximately five miles, the effect of which is to expedite considerably the arrival, departure and smooth flow of aircraft. However, it may be noted that, in practice, air traffic control units use a mixture of procedural and radar control, applying procedural separation vertically and radar separation horizontally.

The Airways

To channel the flow of air traffic and obtain the degree of order necessary to the separation of the aircraft, a network of airways controlled from main air traffic control centres has been established throughout the world. In the United Kingdom, these airways link the main areas of population and connect with the capitals of Europe and the North Atlantic routes as shown in figure 1.

[3]See also footnote on Secondary radar, page 21.

18

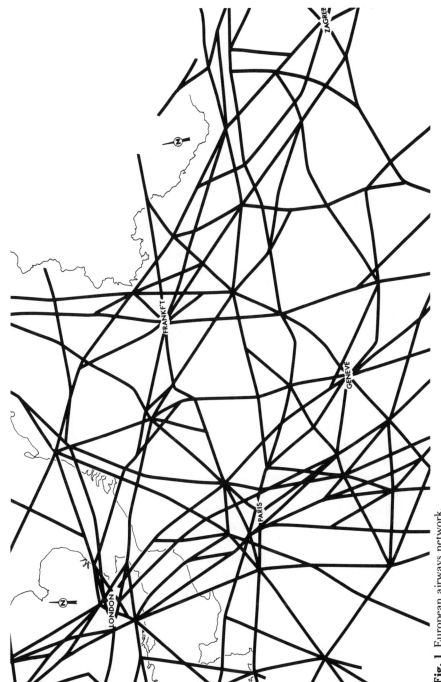

Fig. 1 European airways network.

19

The airways are generally 10 miles wide and are delineated on aeronautical charts by specific colours. They are also identified by signals from ground-based navigational aids such as VOR (very-high-frequency omnirange) radio beacons. The base of the *en-route* portion of the airways is variable, depending on terrain and the need for freedom of movement for certain elements of aviation – light aircraft, etc. – operating below that base.

The controller, then, monitors and organises the flow of aircraft around, through, into or out of this complex network. He is required to separate their flight paths in accordance with the standards previously described and to ensure *whether he is handing an aircraft over to the controller at a destination airport, or to another airways sector, or to an adjacent air traffic control centre, that separation exists and will remain constant.* Those requirements should be borne in mind since they will prove to be of special significance to the central events of this book.

In the above description the controller is handling airways or *en-route* traffic and is therefore located at a main air traffic control centre. To complete the picture and since reference will subsequently be made to controllers' ratings it will be useful to differentiate between this task, which requires its own special training and rating in Area (Airways) Control, and that of the controllers at aerodromes who deal basically with, and are appropriately rated for, Approach Control and Aerodrome Control. The exact duties of these controllers will vary from one location to another, depending primarily on its complexity in relation to the surrounding terminal area, adjacent airports or the volume of traffic with which it has to deal. But, in general, Approach Control accepts inbound traffic from the parent centre controlling the airway and sequences the aircraft, usually using radar marshalling techniques, on to the extended centre line of the runway in use. At this point, and when the pilot has captured the instrument landing system, control is transferred to Aerodrome Control for the final landing (for outbound aircraft the flights are usually so arranged that they do not conflict with the inbound routes). The task of Aerodrome Control is to receive inbound aircraft from Approach Control, ensure that the runway is clear for landing, issue such a clearance and thereafter pass the essential instructions for the taxiing of the aircraft to its parking bay. For outbound aircraft it is the responsibility of Aerodrome Control to obtain an airways clearance from the parent centre, pass this to the aircraft, specify the departure runway, issue taxiing instructions, route the aircraft to that runway and finally issue a take-off clearance when Aerodrome Control is satisfied that all is safe for that operation.

Radar Competence and Controller Training

Both Approach Control and Area Control ratings may be augmented by radar endorsements where controllers have passed the appropriate examinations and completed the requisite experience requirements. Thus, in addition to his grounding in air legislation, I.C.A.O. requirements, meteorology, aircraft performance, Aerodrome and Approach Control procedures, telecommunications, R/T procedures and phraseology, navigation and navigational aids, the radar-qualified controller will be versed in primary and secondary radar[4] and in the radar procedures appropriate either to Area or Approach Control. His instruction will certainly have included simulator training wherein computerised programs of air traffic are used to familiarise the trainee with the problems and techniques of Radar Control.

The training undergone by a contemporary air traffic controller in Yugoslavia largely follows this pattern, for during a typical 12-month course (now extended to 14 – 15 months and including a month of practical training), the curriculum covers the following areas:

Flight (ATC) Control/Theory
Flight Control/Practical Work
Automatic Access and Landing Systems
English Language and English Aviation Phraseology and Terminology
Aviation Meteorology
Navigation and Radio-Navigation
Telecommunications Systems, Equipment and Aeronautical Services
Aviation Rules and Regulations
Airport Infrastructure and Airport Construction
Aerodynamics
Radar Technique
Aeronautical Information Service.

Successful completion of this course establishes the basic qualification for the air traffic controller; and other professional ratings permitting him to undertake Procedural Control or Radar Control responsibilities are obtained 'on-the-job' after appropriate periods of practical training.

---

[4]In primary radar pulses (signals) from a transmitter are reflected back from any object 'struck' by those signals. The reflected signal appears as a 'blip' on the ground controller's radar display, thereby revealing the presence – although neither the altitude nor identity – of the 'target' object. In contrast, secondary radar provides information on location, height and identity but, in order to do so, must interrogate a transponder (receiver-transmitter) installed in the aircraft.

Chapter Two

The foregoing is a drastically condensed description of the air traffic control function and of the background of special knowledge which the controller is required to possess. Brief though this description necessarily is, it can leave no doubt as to the personal capacity essential to the task or the weight of responsibility it imposes, not only on the controller – the man at the console – but also on the employer authority; for, heavily imbued as they are with social and moral implications, the manner in which these professional obligations are discharged is dramatically influenced by the national political structure and the type and quality of the administration which serves it. It is a fact, however, that for those reasons and the attendant economic, technological and human factors which complicate them further, no air traffic control system anywhere in the world can claim to be altogether satisfactory, while in many systems the scale of shortcomings is of intimidating proportion.

At best, controllers may serve enlightened and fortunate administrations as highly organised professionals in locations which are equipped to the very highest order with every necessary modern device. They may, nevertheless, be preoccupied with such vexatious disputes as that arising from the pressure (inspired by considerations of nationalism rather than safety) for bi-lingual ATC control in the French-speaking Province of Quebec,[1] or with any other of the inevitable conflicts between management and employees which are common in every sphere of commerce or industry. At worst however, they may find themselves denied professional recognition and equated with train controllers or telecommunications technicians; they may be under military authority, with all that is implied by that phrase, or, in the face of any proposed protest by strike action, be threatened with prosecution or by replacement with military personnel: and of similar import, they may be at work for long hours at low pay, in primitive conditions dictated by the poverty of a national economy, enjoined, nevertheless, to carry out their task as the guardians of aviation safety

---

[1]Legalised despite considerable opposition within civil aviation on the grounds of the patent threat to air safety, and despite such precedents as the collision of 25 February 1960. This accident, involving a United States Navy DC6 and a DC3 operated by Real Aerovias of Brazil, claimed a total of sixty-one lives. The controller made all transmissions to both aircraft using the same radio frequency for both but using Portuguese for communication with the Real DC3 and English for the Navy DC6. All evidence proved conclusively that the use of two different languages to conduct Rio air traffic control was a major, if not the primary cause, of the accident. See also pages 70 and 154–5 for relevance to Zagreb.

with old or deficient equipment, or with ill-assorted radar or electronic systems. And all this, quite possibly, under the unpropitious circumstances of inadequate organisational or airspace management.

And lastly, there are the human factors mentioned above.[2] For all of these impediments may at any time be reinforced by mankind's own and very ordinary weaknesses, and by the nature of his response to the manifold stresses of life and events. The effects of these pressures will be examined more closely in the appropriate chapters of this book: it suffices now that some, at least, of these elements were reflected in microcosm on Friday 10 September 1976 at the Zagreb Regional Flight Control Centre.

[2]See also Weston, Richard 'Human factors in air traffic control' in Hurst, Ronald and Hurst, Leslie eds. (1982) *Pilot Error*. 2nd edn. London: Granada.

Chapter Three

The air traffic control centre at Zagreb Airport is part of that service guarding a major crossroads of the busy European air routes. Great numbers of aircraft overfly this region at the rate of scores every hour: a heavy and demanding traffic *en route* for Europe or south and eastward to Greece, Turkey and the Mediterranean and beyond.

Five airways converge on the Zagreb VOR beacon, which identifies the junction: in turn, each airway is identified on aeronautical charts and for briefing purposes and the allocation of flight levels and in operations by the radio beacons sited along its route. These beacons serve also as reporting points for the aircraft, enabling the controller to check the progress of each flight and, where necessary, to make the deployments which ensure separation with the airspace.

The illustration (figure 2) shows that the airways can be seen as a constellation in which the top is formed of a conjoined V with Zagreb at the apex. This point is linked by the stem of an inverted Y, the whole figure some 200 miles in length lying almost due north-west to south-east in line with the orientation of the Adriatic coast.

A further airway passes through Zagreb from east to west, forming a transverse bar across the main figure.

The northerly extremities of the V are the beacons of Graz and Klagenfurt in Austria: the transverse airway is signalled on the western side by the beacons at Ilirska Bistrica, between Fiume and Trieste, and Metlika, midway between Ilirska Bistrica and Zagreb; and on the eastern side by the beacon at Nasice, *en route* to Belgrade.

The stem of the inverted Y runs from Zagreb to Kostajnica, about 80 miles to the south-east. Here the arms of the Y diverge, one airway reaching out to Sarajevo, the extreme south-easterly point of the constellation, and the other, to Split. Yet another airway, subsidiary to the constellation, bypasses Zagreb branching north-west from Kostajnica to Metlika and running beyond to diverge at Dolsko for Klagenfurt and Filah.

This then is the regional pattern of the airways. In terms of aircraft speed, none of the distances between these reporting points is great, and in fact, an airliner would require less than 40 minutes to travel the entire length of the airways from the Austrian border to Sarajevo or Split. But there is, of course,

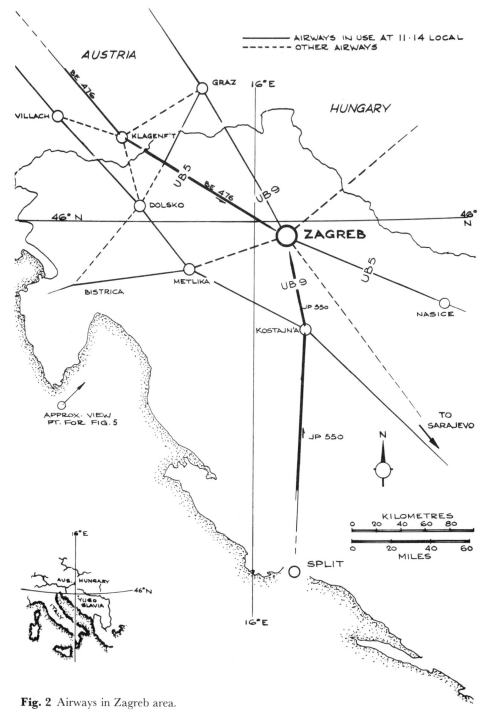

AIRWAYS IN USE AT 11·14 LOCAL
OTHER AIRWAYS

AUSTRIA

BE 476

GRAZ   16°E

HUNGARY

VILLACH

KLAGENF'T

UB 5

BE 476

UB 9

46° N   DOLSKO

46° N

ZAGREB

BISTRICA   METLIKA

UB 9   UB 5

NASICE

JP 550

KOSTAJN'A

APPROX. VIEW
PT. FOR FIG. 5

TO
SARAJEVO

JP 550

N

KILOMETRES
0   20   40   60   80

0   20   40   60
MILES

SPLIT

16°E

AUS. HUNGARY

YUGO
SLAVIA

ITALY

46°N

16°E

Fig. 2 Airways in Zagreb area.

the pressing multitude of aircraft; each contending for airspace, flight levels and above all, for the attention of the controller and the guardianship he represents.

On the morning of 10 September that guardianship lay in the hands of eight men: the five air traffic controllers responsible for the upper and middle airspace sections[1] invigilated from Zagreb; the head of the duty shift; and the representative heads of the flight control service and the district flight control. It is timely now to explain the most vulnerable members of this group and to note their collective experience of the hazards of their profession.

The Morning Shift, 10 September

With the exception of 43-year-old Julije Dajcic, the chief of shift, the duty controllers were all relatively young men.

Dajcic, born at Pola, was described as an irascible and hard-driving supervisor, yet one should mark the very real anxieties which doubtless contributed to that image; for the difficulties of a professional life so clearly unforgiving of laxity or error were further compounded by a rapidly expanding air traffic in the region and a chronic shortage of trained personnel and technical equipment. Thus, the Zagreb centre which had been overflown by some 700 000 aircraft in the previous 5 years had now become the second busiest aircraft crossroads in Europe – an eminence which, not surprisingly in that time, had generated its total of 32 air-miss incidents, the dismissal of two controllers for carelessness and lack of training and a recorded 166 critical remarks in the airport's complaints book.

Dajcic's own professional experience was considerable since he had obtained his area radar rating in 1963 – a seniority which could only have added fuel to his frustration at the shortcomings of junior colleagues in matters of performance or acceptable behaviour. Such failures were by no means insignificant, for of the five men on the duty shift, only one had escaped previous censure for offences ranging from late arrival and unauthorised absence from the station, to that of actually endangering aircraft by neglect of procedural discipline.

So baldly stated, the sum of these derelictions appears wholly to condemn an unusually irresponsible and unpromising quartet: but it is fair to say that this is perhaps inevitable in a record which must dwell on omissions rather than achievements and on those qualities reflected in the behaviour of this group at that time, and in the aftermath of a major disaster.

Similarly the charge of endangering aircraft should be seen in the perspective of comparative experience and of the great numbers of aircraft which had been passed safely through the region. It is also salutary to remind readers that, given the complexities of the task, there can be few controllers who in the course of

[1]There was, of course, a lower airspace sector. Its personnel were not involved in this event.

their working lives have not been rescued from potentially disastrous situations by the timely intervention of pilots or colleagues.[2]

In the 7 days before 10 September, Julije Dajcic had a 48-hour stand-down covering the 4th and 5th. He had worked on the 6th as chief of shift from 07.00 to 19.00 hours local time and again on the 7th from 19.00 to 07.00 hours on the 8th. His next spell of duty was once more in the capacity of chief of shift from 07.00 hours on the 10th.

The oldest of his subordinates on that occasion was Nenad Tepes aged 34, born in 1943 at Zagreb. Tepes, on duty as an assistant controller, held a final approach controller's licence and an area radar licence which he had obtained in 1975. Like Dajcic, he had stood down on the 4th and 5th and worked from 07.00 to 19.00 hours on the 6th. He had not worked on the 7th, had put in the 07.00 to 19.00 hours shift on the 8th, stood down on the 9th and resumed his duties with the morning shift at 07.00 hours on the 10th.

Next came Mladen Hochberger aged 30, born in Zagreb in 1946. Hochberger had qualified as a licensed radar controller in May 1976 and in the week before joining the shift on 10 September, had had four full days off duty – on the 4th and 5th and on the 7th and 8th.

Bojan Erjavec, 30, was born in Belgrade in 1947. He had obtained a district procedural controller's licence in March 1976 and had been on holiday during the 7 days prior to the 10th. Erjavec was a flight controller and was assisted during his watch by Gradimir Pelin aged 29, and also born in Belgrade. Pelin's licence as a radar controller dated from April 1976; his duty and stand-down periods for the relevant 7 days were similar to those of Hochberger.

The muster of the morning shift was completed by Gradimir Tasic, its youngest member, aged 28, born in Nece, Serbia and educated in Belgrade. He had passed a district radar examination on 26 March 1976 and was the only person in the centre who had worked on the preceding day. He had also worked the same shift from 07.00 to 19.00 hours on the 8th, stood down on the 7th and worked that shift again on the 6th, following stand-down on the 4th and 5th. The chart in figure 3 shows the pattern of duty and stand-down days for each of the controllers. It will be seen that only Dajcic, with a 12-hour shift between the 7th and 8th remotely approached the sustained attendance of Gradimir Tasic during this period. Similarly it may be noted that between the 8th and 10th, Tasic's two off-duty periods were of no more than 12 hours each, while his colleagues, in contrast, achieved a full 24 hours between each duty spell.

It was Tasic who was to become forever identified with and branded by the events of that morning and who, for that reason, requires a more detailed

---

[2]In reality, however, a high proportion of errors within the ATC environment are never reported, for there is a great degree of sympathetic appreciation of the mutual problems of pilot and controller. It is customary, as a result, for errors arising during heavy, complex or difficult traffic situations to be dismissed by a brief apology, and it is, in fact, only when emotions are stirred by an actual or potential hazard that an official report is made. See Martin, Philip 'Air traffic control factors' in Hurst, Ronald *ed.* (1976) *Pilot Error.* 1st edn. London: Granada.

| | DAY No. | | | | | | | HRS. WORKED DURING 5 DAYS |
	1	2	3	4	5	6	7	
DAJCIC			X	X			X	///////// 28¼
TASIC			X		X	X	X	///////////// 40¼
TEPES			X		X		X	///////// 28¼
HOCHBERGER			X		X		X	///////// 28¼
ERJAVEC							X	/ 4¼
PELIN			X		X		X	///////// 28¼

DUTY PERIODS :- X

Fig. 3 Controllers' duty pattern.

scrutiny here. He was to be described as 'a small, thin man with a pale complexion, dark short hair and dark eyes; a quiet man who was difficult to talk to and who kept himself very much to himself. He was, apparently, an excellent carpenter who made his own furniture, and his colleagues would visit his home to admire his handiwork.'[3]

Other views endorse Tasic's reserve while blocking-in the most readily observable facets of his character. Thus 'Tasic comes across as a more complex personality (than any of his colleagues). Introverted, quiet, self-confident, highly-strung – all these adjectives have been used to describe him. Essentially he is a loner, a man whose main interest is work.'[4]

Plainly, no man is to be summed in such brief comment and Tasic's outward personality undoubtedly concealed its more profound shades. Certainly, following his period of National Service in the Army, he had been sufficiently imaginative and sufficiently self-confident to aspire to a most demanding profession: and certainly he had impressed his selectors and had been accepted in 1971 as a suitable candidate for the ATC Training Centre in Belgrade. Subsequently, after one year's training, he had been posted to the Zagreb District Control and at that station had gained, first, his procedural controller's licence – in 1973 – and thereafter, as noted, his licence to operate as a radar controller.

His service, it is true, had not been without its occasional lapses of discipline, yet in the course of that service he had also won a considerable regard for his competence;[5] and this despite a social and domestic environment which must have challenged every instinct for improvement.

[3]David Liddiment, researcher for Granada Television programme 'Collision Course', August 1977. This programme was transmitted on 20 February 1979.

[4]*Sunday Times* 17 April 1977. The correspondent, Michael Dobbs, drew a further contrast: 'Even his clothes set him apart from the others. He wears an open-neck shirt and a crumpled blue uniform while his colleagues appear...in well-cut suits.'

[5]The transgressions are listed in the indictment (see page 88); cf. with endorsement of Tasic's professional capacity, page 96.

It was not merely that both Tasic and his wife Slavitsa were Serbs constrained by the fact of his location at Zagreb to adjust to the atmosphere of a Croatian community; for despite the undoubted nationalism of each of Yugoslavia's seven republics, the ability of its people to live and work in harmony is demonstrated daily. Such an integration might well have been eased had the pair found themselves in acceptable personal circumstances; but in the event, few might have been prepared to settle for long for the desolate makeshift which Tasic and his wife had been obliged to call their home. For Slavitsa, indeed, marooned alone during her husband's working hours it became impossible: soon she would return to Belgrade.

Meanwhile, it was another facet which distinguished him from his colleagues, for, while their homes were in the city and while they enjoyed its life and found pleasure in each other's company,[6] the Tasics, as a temporary solution to their accommodation problem, had been allotted an abandoned radio-equipment hut within the airport area, one kilometre from the nearest building and some two kilometres from the ATC centre.

Surrounded, as this remote dwelling was, by the stark masts of the airport's radio transmitters and fenced by wire into a bare compound, the site could not have made for anything but a dour and lonely existence. In that situation there was logic and continuity in the preoccupation with carpentry, while for Slavitsa, there were the demands of her baby; but the frustrations of their predicament – for the appointment to the Zagreb air traffic control centre had not, it seemed, included the provision of more suitable quarters, nor had Tasic been able to improve matters by his own efforts – were not to be borne without some struggle for escape.

It had taken time and dogged persistence to achieve that goal: but it seemed to them now that escape, in the shape of an impending transfer to the Belgrade air traffic control staff, was truly at hand and that there could at last be an end to the wistful pilgrimage between that bleak hut and the Zagreb control centre.

Tasic was not to know that that end had already come: and that, on this morning of Friday 10 September, he had made that journey for the last time.

Now, at 7.0 a.m.,[7] he would join the others for the morning shift. He would act as controller for the upper airspace sector – and thus be responsible for aircraft flying at or above 31 000 feet – until 9 a.m., take a one-hour break and then return to the upper sector console in the role of assistant controller until noon.

In this capacity his major task would be to smooth the work of the controller by undertaking any necessary co-ordination with neighbouring control centres and, according to the duty schedule, he would serve two controllers in this way: first, Mladen Hochberger and second, Nenad Tepes who would relieve Hochberger at 11 a.m.

[6]Description of the Zagreb air traffic controllers by Michael Dobbs, *Sunday Times* 17 April 1977.
[7]Shifts are described in local time. All other times stated are G.M.T.

There would be a similar change of personnel at the middle sector position which monitored the airspace between 25 000 and 31 000 feet. Bojan Erjavec would become controller at 10.00 a.m.; and at 11.00 a.m. his own assistant would be replaced by Gradimir Pelin.

Thus, the men of the morning shift in the Zagreb control centre took their places before the radar screens, the microphones, through which they would speak to the aircraft in their sectors, and the banks of telephones linking the control network. About them was the special setting of their task: the subdued lighting pierced by the brighter tracery of each radar plot, the quiet of the room and, against the right-hand wall, the conjoined, side by side, array of the lower, middle and upper control stations each with its position for the radar controller, the procedural controller and the assistant. Bedevilled by the lack of local accommodation however, and in consequence chronically short-staffed, the Zagreb administration could afford no such luxury as a three-man watch for each sector – hence the current deployment of paired men who, with the exception of Erjavec who lacked a radar rating, could handle either procedural or radar control as required.

Against the left-hand wall stood the terminal sector stations for aircraft arriving or taking-off from the airport. Places for the military controllers were located at the far end of the room while at the centre were the desks of the flight data girls who would process information from the flight plan for every aircraft scheduled to fly through the region. This information would subsequently appear on the respective procedural controller's console in the form of a flight progress 'slip' detailing the aircraft's projected time of arrival in the sector and its airspeed, altitude and routeing; it would, therefore, enable the controller to review the future traffic situation and prepare to marshal his traffic accordingly.

By the main doors, flanked by television screens displaying sector and meteorological information and strategically placed to take in the entire scene, sat Dajcic, chief of shift. He looked on as the men of the outgoing watch briefed their replacements; as headsets and microphones were adjusted and as the first quiet calls began to be heard; as the staff settled into another day's routine; and as Tasic began his third consecutive day of duty.

BE476: The Flight Crew

British Airways' Flight BE476 which took off from London Airport (Heathrow) at 08.32 (G.M.T.) that morning for a scheduled three-and-a-half-hour direct flight to Istanbul, carried 54 passengers and crew of nine, of whom six were cabin staff.

The aircraft, registered as G-AWZT and hence designated by its call sign Zulu Tango, was a Trident 3B, a 'stretched' version of the original short-haul aeroplane designed by Britain's Hawker-Siddeley Company in 1959.

The Trident 3 variant was usually fitted out by British Airways to carry up to 140 passengers seated six abreast, and, with its three Rolls Royce Spey turbofan engines clustered at the tail (and augmented for take-off by a small booster engine mounted above the rear fuselage), could cruise for 3000 miles at more than 500 miles per hour.

The route which BE476 would follow would take the aircraft on a southeasterly course from London. It would cross the English coast at Dover and overfly Brussels, Munich and Klagenfurt, on the southern border of Austria, and continue via Zagreb and Belgrade, to touch down at Istanbul at 12.45 local time.

BE476 was under the command of one of British Airways' senior pilots, Captain Dennis Victor Tann, aged 44, who had joined the airline – then British European Airways – in 1957. A married man with three teenage children, Tann held a current airline transport pilot's licence and had amassed a total of 8855.6 flying hours with British Airways. Of this total, some 400 hours had been on the Trident 3.

In the previous 28 days he had logged 33 flying hours, the last three days accounting for 2.12 hours – figures which may well be misleading for those unacquainted with the routine of the line pilot. In fact, Captain Tann had begun the first day of the past week at 05.50 hours by flying from Glasgow to Heathrow and thence to Edinburgh, returning to Heathrow and going off duty at 14.17 hours. His flying time for that day had been 3.14 hours. He had been at his home at Iver, Buckinghamshire, off duty on days 2 and 3 and on a six-hour standby at home on both days 4 and 5. Day 6 saw him at the airport once more; he had gone on duty at 07.30 and had flown to Brussels, returning to Heathrow and signing off at 12.38. Day 7 was Friday 10 September; and on that day he took command of BE476 with 29-year-old Brian Edward Helm of Ash Vale near Aldershot, as his first officer. Helm, too, was married with two sons aged three and four. He had joined British Airways in 1969, possessed an airline transport pilot's licence and had acquired 3414.3 flying hours, 1592.5 of these on Trident 3 aircraft.

The previous week had not been exacting. He had been off duty for the first three days and on standby at Heathrow on day 4. He had reported for duty at 05.40 on day 5 and operated a flight to Paris and back to Heathrow, recording a flying time of 1.55 hours. He had been on standby at home on day 6 between 09.55 and 15.55 and was now, on day 7, rostered as P2 to Captain Tann.

The youngest member of the flight crew, 24-year-old Martin Jonathan Flint, of Manchester, had been with British Airways since 1973 and held a senior commercial pilot's licence. He had acquired 1497.7 hours of flying time with the airline almost all of which had been on the Trident.

And like Brian Helm, Acting First Officer Flint could look back on a somewhat undemanding week. He had flown to Rome and returned to Heathrow on day 1, clocking up 4.49 hours of flight time and going off duty at 13.36. He had gone on duty at 09.40 on day 2 and had added a further

2.14 hours to his flight time during the return trip from Heathrow to Edinburgh; thereafter, he had signed off at 14.15 and had followed a two-day stand-down with two days of compassionate leave in order to visit his sick father before reporting at 07.20 on day 7 for deployment to BE476 and the privilege, for this young man with a passion for aviation, of occupying the third man's perch – the engineer's jump seat behind the two pilots.

In the bright cockpit now he immersed himself in the routine of the climbing aircraft, in the interchange of dialogue and movement between the captain and his first officer. Behind them the cabin crew moved forward to begin the first of their own in-flight tasks.

BE476: Cabin Staff and Passengers

In terms of their airline service, the four stewardesses and two stewards who comprised the cabin staff represented a remarkably stable group. Chief Stewardess Anne Whalley, 32, had come to British Airways in 1966: Rona Goddard-Crawley, 31, who was married to an air steward, had also served for ten years, while Stewardess Jennifer Munday, who had only achieved her ambition to fly a mere four months earlier, had in fact joined the airline in 1969.[8] Jennifer was also married. Stewardess Ruth Pedersen, from Denmark, aged 26, had served British Airways for four years. Chief Steward Lawrence Joseph O'Keefe, 30, had joined in 1970, while Chief Steward David John Crook, 33, married, had begun his airline service in 1965.

The cabin staff were thus an experienced and competent unit and on this flight, without the pressures generated by a full aircraft, they could deal unhurriedly with the needs of its 54 passengers – an invariable mix of nationalities which included Britons, Turks, Anglo-Turks, Turkish Cypriots, Australians, South Africans, Saudis, Americans and one stateless person.[9]

As varied, too, were the motives which had brought them aboard the Trident. Mehmet Aksoy, 54, was flying home to his wife and family in Nicosia after visiting his son in London:[10] 19-year-old Martin Heffer, a deck officer cadet with the British Petroleum Fleet would be joining his ship British Loyalty at Istanbul after a six-week leave. With him was his shipmate, seaman Robert Siner, 47, a married man with two children, who had once attempted to give up the sea to please his wife. The effort had not kept him shorebound in Liverpool for long. An unsuccessful stint as a window cleaner had decided that struggle for him and he sat now beside young Heffer, a wry and resigned companion for the 'boy who lived for the sea'.

[8]The dedication and sense of vocation so commonly found in the air transport industry is exemplified by the following quotations: On Anne Whalley: 'she really loved the life. I think she stayed single because of her job.' (Miss Whalley's father during *Daily Mail* interview, 11 September 1976.)

On Jennifer Munday: 'She was thrilled; in the last year it had really got into her blood. She would give up days off to go on escort flight duty.' (Interview with Mr Barry Munday, *Daily Mail* 16 September.)

Patrick Browne, the 41-year-old father of three, who ran his own computer business, had been called to Istanbul to carry out maintenance on the computer on the Bosphorus Straits bridge: Company Secretary Derek Hoare, 25, looked forward to a two-day business round in Istanbul, while Turkish-born Hilmi Halmamzi, 35, a mechanic who had settled in England had briefly left his wife and two daughters at home in London in order to visit his mother and sister in Turkey.

Elspeth Brown, 22, a social worker and Nigel Balin...were taking a three-week holiday after graduating from the University of York. Kathleen Rouse, 20, who had just completed her second year as a biochemistry student at the University of Manchester, was on her way to Ankara. Kathleen would be welcomed by her friend, Ali Guren Guchdemn who had studied with her and had now returned to Turkey for his military service.

Similar backgrounds, similar emotions and similar ambitions remained concealed by the printed names on the passenger list for Flight BE476. Absorbed in newspapers and magazines, in the eager chatter of those beginning a special journey, or simply by the sunlit cloud carpet spreading below, the passengers settled down as the Trident completed its climb, levelled out at cruising altitude and took up its course across Europe.

JP550

Filled almost to its maximum capacity the DC9 aircraft which took off from Split at 10.48 that morning carried 108 passengers – 107 West Germans and one Yugoslav – whose immediate purposes were considerably less complex. They had enjoyed their holiday on the islands of Huan and Brac on the Dalmatian coast and had boarded Flight JP550 for their homeward flight to Cologne.

The aircraft, registered YU-AJR and operated by Inex-Adria Aviopromet, the Yugoslav charter airline, was one of its fleet of four McDonnell Douglas DC9 Series 32s. Two aft-mounted Pratt and Whitney turbo-fan engines enabled it to cruise at around 500 m.p.h. over a range of 1100 miles; a performance which had earned the machine its place as one of the most popular types in airline service.

For Captain Joze Krumpak, 51, the flight deck of that aeroplane was a very familiar work place. He had served with the airline since 1962 and had acquired some 3250 of his total of 10157 flying hours on this aircraft. Ninety-four flying

---

[9]It is of interest that different newspapers reported the passengers thus: 'The Trident passenger list: included 20 Britons, 16 Turks, 2 Anglo-Turks, 3 Turkish Cypriots, 6 Australians, 6 South Africans, 5 Saudis, 3 Americans, 1 Dane and 1 stateless person.' *British Airways News* 17 September 1976.

[10]Mehmet Aksoy was described as being 54 by the *Daily Express* of 11 Sept. but reported in the *Observer* 12 Sept. and in the *Evening News* 11 Sept. as 62.

hours had been attained in the past month, 18 of them within the past 3 days of an extremely industrious week.

He had been lucky enough to spend day 1 at home. On day 2 he had flown Dubrovnic–Stuttgart–Dubrovnic and from Dubrovnic to Hamburg. On day 3 he had flown Hamburg–Tivat–Lubliana–Tunis and back to Lubliana and had gone home on day 4, only to have his stand-down interrupted by a call to fly to Cologne, back to Pula and outward once more to Hamburg. How far these flights extended into day 6 is not clear: but on the morning of day 7, 10 September, he had already made the flight from Hamburg to Split, where he had taken command of JP550.

Life had been only a little less hectic for his second pilot, 29-year-old Dusan Ivanus ('Branko'). Ivanus, with 2951 hours to his credit, had joined the company only two months earlier but in the past 28 days had put in 84 hours, 11 of them on the previous 3 days. More fortunate than his captain, he had flown on day 1 of the last week (Dubrovnic–Bristol. Dubrovnic–Lubliana) but had then spent the next three days at home. On day 5 he had flown Pula–Dusseldorf–Pula–Hamburg, on day 6, from Hamburg to Pula and back to Hamburg, and on the morning of day 7, from Hamburg to Split.

He had come now to his second task of day 7 and once more, to the small kingdom and deft teamwork of the DC9's flight deck; and he too, watched his captain, vigilant for Joze Krumpak's quiet commands as JP550 began its climb over the Split VOR beacon.

Within the passenger cabin it was clear to Lidija Ofentavsek that on this trip, the flight attendants were in for a very busy time indeed. At 29, Vili Ofentavsek, born in Maribor, was one of Adria's chief stewardesses and with her own tiny team, 23-year-old Mojca ('Franc') Sila and 24-year-old Jelka ('Vinko') Zagar, now faced the prospect of coping with about three dozen passengers each within the next couple of hours. They would also need to remember the two men in the cockpit: they would certainly need to be brisk, then, if they were to be finished with the meals and drinks and innumerable requests for service in time for the landing approach. . .In the course of that flight the rows of seated figures would resolve themselves into recognisable faces and personalities; into the strange and yet everyday mix of human beings and human preoccupations typifying every plane load. Like those of the young couple in *that* row, for instance. . .

Wilfried Knaut relaxed in his seat next to his fiancée Barbel von Chrzanowski. It had been a 'lively holiday' at Huan, he had written to his parents in Cologne/Rath and he'd ended with every postcard writer's anodyne: 'Tell you all the details when I get back.'

But 'back' was an equally lively scene, since Wilfried who worked as an export executive for a firm in Stuttgart – and consequently maintained an apartment in that city – commuted back to Cologne almost every weekend to see his parents and to meet Barbel, who was a nurse at the University Hospital. They would be married in the spring and would move into the house he had

34

begun to build for all of them at Hennef.

The two ladies who looked about them and out at the DC9's gently flexing wing were old friends. Adele Longhorst, a 58-year-old sales assistant for a Cologne textile wholesaler, and Lotte Bergman, from Porz-Zundorf, habitually spent their holidays together, usually in two two-week breaks each year. This trip, therefore, had followed a visit to Greece, in May. Content, they chatted amiably while about them, other passengers came momentarily into relief.

It must certainly have been a welcome break for Rolf and Helga Bauer, now returning to their home in Zollstock. At 50, Dr Bauer was in the process of vigorously consolidating a successful dental practice and of benefiting from the results. They were getting away more often now and there were so many exciting things in the future...

In their own seats, Gunter and Emmi Ziemer radiated the calm presence of an elderly couple. Soon after the Second World War had ended, the Ziemers had found themselves a house in Altstadt, North Cologne, and Gunter a job in the insurance business. The job was a long way behind him now, and Emmi a placid 70; but retirement had at least brought the opportunity for this kind of adventure and it could have been nothing less for the two quiet people who were still in that same house on Maybach Street.

Cologne and its environs in fact represented home for most of the passengers: now, while JP550 lifted through the clear sky, they subsided into the anonymous harmony of the first hour aloft, busied themselves with a score of trivial adjustments and pursuits: and here and there, comforted restless children.

Chapter Four

By nine o'clock the traffic in the region had begun to impose a steady demand on the control staff in the Zagreb Centre, now dividing their time between telephone requests for co-ordination, radar monitoring and direct contacts with aircraft in transit.

At 09.36 one such request became the subject of a telephone conversation between the controller of the Zagreb lower airspace sector and Split Approach Control. Its purpose was to clear the outward flight pattern for JP550, whose crew had filed a flight plan for an altitude of 31 000 feet.[1]

Accordingly, the controllers agreed that the aircraft should climb to 12 000 feet over the Split VOR beacon; it would then switch to Zagreb Regional Flight Control for permission to continue its climb to 19 000 feet while homing on the Kostajnica beacon.

Each step in the ascent may be followed from the record, presented here against the critical timebase. It should be noted, however, that the transmissions between the controller and other aircraft in the sector are equally important since they indicate the build-up of traffic and the consequent work load for the controller in terms of information to be absorbed, mentally and physically cross-referenced, and used as a basis for subsequent control guidance. Nor can the very short intervals – of seconds only – between each transmission fail to imply the pressures for urgency and clarity under which that guidance must be given.

At 09.54 and 49 seconds, then, JP550 – using the airline call sign 'Adria' – made its first contact with the controller of the lower sector on a frequency of 124.6 MHz.

09.54 49"	JP550	*Dobar dan (Good-day) Zagreb, Adria 550 crossing 130,climbing 180, heading Kostajnica.*
09.55 01"	Zagreb	*Roger, recleared 240, Adria 550.*
	JP550	*Recleared 240.*

In this exchange Captain Krumpak confirmed that he was passing through the 13 000 foot level and climbing for 18 000 at Kostajnica. The message was

---

[1]Note: aircraft en-route altitudes are normally measured in thousands of feet and referred to as 'flight levels' – thus, 18 000 feet becomes flight level 180: 24 000 becomes 240, etc.

acknowledged by the controller, who then cleared JP550 to 24 000 feet, received the captain's reply, and turned his attention to another Inex aircraft, JP548, *en route* from Split to Nuremburg.

09.55 22″	Zagreb	*Adria 548, level check?*
	JP548	*Out of 190 (climbing from Flight Level 190).*
09.55 26″	Zagreb	*Thank you. . .*

Time to call 550:

| 09.55 50″ | Zagreb | *Adria 550, recleared 260, call passing 220.* |

He waited. Within 12 seconds he called the aircraft again.

| 09.56 02″ | Zagreb | *Adria 550, Zagreb. . .* |

Now the reply came at once:

| 09.56 06″ | JP550 | *Cleared 260 and call you passing 240, do you read me?* |

Two-forty, noted the controller: I told him 220. Patiently he repeated the instruction:

09.56 12″	Zagreb	*Call me passing 220.*
09.56 15″	JP550	*I will call you passing 220.*

For the next 38 seconds the controller worked the aircraft LZ696 and JP548:

09.56 42″	LZ696	*Zagreb, Balkan 696, dobar dan, brakto (Good-day brother).*
	Zagreb	*Dobar dan 696, go ahead. . .*
09.56 47″	LZ696	*Overhead Tuna at 55, level 230, estimating Split 05. Would you like Squawk?*

At 09.55 LZ696 was overhead Tuna at 23 000 feet and estimating his arrival over Split at 10.05. As noted in chapter 1 the activation of the Squawk from the aircraft transmitter would bring an alphanumeric code group (selected by the controller) and a height read-out alongside the blip which represented that aircraft on the controller's radar screen; the aircraft would thus be positively identified. Yes, let him Squawk, decided the controller. He called the aircraft:

09.56 54″	Zagreb	*Squawk Alpha 7200, call Split*
	LZ696	*7200 is on. . .*

The number glowed beside the firefly dot that was LZ696. Satisfied, the controller took in the rest of the traffic plot, expertly calculating priorities. 548, he registered, was approaching the upper limit of the sector and would need to go over to the middle sector frequency. He called with the instruction:

09.58 12″	Zagreb	*Adria 548, contact Zagreb 135.8*
		Good-day. . .
09.58 17″	JP548	*Adria 135.8 548, Goodbye. . .*

The watchful eyes searched the radar screen for JP550, singling out one point of light among the constellation. Time to check; the aircraft should have made considerable height by now.

| 09.59 53″ | Zagreb | *Adria 550, level check?* |
| | JP550 | *550 passing 183 –* |

The controller's acknowledgement immediately triggered calls from KLM283 reporting its height and from Echo Whiskey, estimating its arrival over the Nasice beacon; and, as if the man at the console had orchestrated a perfectly judged trapeze handover, both aircraft were cleared as JP550 made its farewell call to lower sector and received permission to change frequency.

10.02 44″	JP550	*Zagreb, Adria 550 passing 220*
	Zagreb	*Zagreb, 135.8, Good-day –*
10.02 50″	JP550	*Goodbye*

It had taken just eight minutes for JP550 to pass through the lower sector and in that time the controller had dealt with five aircraft. In the cockpit of 550 now, a hand reached for the frequency control and both crewmen waited for the break in the flow of radio transmissions which would enable them to announce their entry into the middle sector.

It has never been satisfactorily established just *why* the Middle Sector controllers had *not* been provided with a flight progress slip for JP550, although Zagreb ACC (Area Control Centre) regulations stipulated, in accordance with internationally accepted practice, that such strips should be prepared in advance for all sectors through which the aircraft intends to pass.

The omission is noted in the official report by the Yugoslav Federal Civil Aviation Administration but is not held to have prejudiced air safety in the Middle Sector since it is affirmed that the handover of the Inex Adria aircraft JP550 from the Lower Sector to the Middle Sector was effected in good time. This transfer was carried out on the basis of mutual agreement between the controllers and the passing of a slip bearing flight information.

It is to be assumed, then, that following JP550's 'Goodbye', the lower sector controller picked out that aircraft's slip from the rack before him and passed it to either Erjavec or Pelin in the adjacent position. He would complete the transfer by offering all necessary information; and incidentally introduce into the middle sector an aircraft about which its controllers, busily marshalling their own correctly documented traffic, had had no previous notice.

Chapter Five

By 09.43 the Trident had completed an uneventful one and a half hours of flight time and had reported its passage over Munich. Captain Tann's next call would be to Vienna Control confirming his E.T.A. over the Villach and Klagenfurt VORs and, until it was necessary to make those transmissions, there was time for that inconsequential and largely insignificant chatter which intersperses all flight deck conversation. It is perhaps necessary to stress that term 'largely insignificant' for the lay reader, and to explain that – fortunately – the handling of an aircraft in flight is not carried out in a state of continual drama. The prosaic and even trivial nature of the dialogue here is therefore no more than typical of that to be heard at times in the cockpit of any airliner and indeed, betokens the calm of experienced crew members wholly satisfied with and regularly satisfying themselves about the progress of the flight and the smooth functioning of their machine.

Relaxation aboard BE476, therefore, came in the form of a newspaper crossword puzzle and an idle question from the First Officer. . .

'Can I have a word "exeat", meaning "to leave"?'
Captain Tann thought for a while. Aloud he said slowly, 'E,x. . .'.
Helm interrupted him.
'E,x,e,a,t', he said. 'Exeat, or something meaning "leave".'
'Yeah,' said Tann, 'it might be. I don't know; er, sounds logical, doesn't it?'
'It does', replied Helm.
'It'll be exeat, or something like that', mused Tann.
'Yeah, exeat sounds feasible', Helm said.
Tann asked, 'What letters have you got in it?'
'I've got. . .it says, "leave the river at ex".'
'The river ex; e, x?' Tann was puzzled.
'That's. . .18, leaves the river. . .', repeated Helm.
Tann wanted to know if Helm had any letters at all to give him a clue.
'First letter's "e"', said Helm.
'Last letter', – Tann thought about it.
'Third letter "e"', continued Helm. . .
Tann agreed. 'Third letter's "e", yeah', he opined.

Stretching for miles behind, the Trident's vapour trails gleamed in the sun. They waited for the next checkpoint and let the crossword serve its purpose: a non-committal bridge between rank and age, a safeguard against the boredom of occasional inactivity: and a contribution to the tranquil atmosphere of the flight deck.

Chapter Six

In the Zagreb Control Centre, Bojan Erjavec, halfway through his shift as controller of the middle sector, was about to call Olympic Airways 187.

10.02 52″	Zagreb	*Report Zagreb, Graz next, Squawk* *Alfa 2500*

OA187 confirmed that its identifying Squawk code of A2500 was on, that it was flying to Graz; and signed off.

OA187   *2500 is, Zagreb, Graz next, Olympic 187.*

Fifteen seconds later, at 10.03 07″, JP548 called in, reporting its position. *Adria 548 approaching Kostajnica, reaching, maintaining 280, Zagreb 08.* Zagreb acknowledged the fact that the aircraft was flying at a height of 28 000 feet, and expected to reach Zagreb at 10.08. The pilot was told to report in at that time and to fly on to Graz. JP548 confirmed and signed off.

Thirty-one seconds after bidding farewell to the lower sector controller, JP550 made its first contact with the middle sector.

10.03 21″	JP550	*Dobar dan Zagreb (good-day), Adria 550 crossing 225,* *climbing 260.*
10.03 28″	Zagreb	*550, good-morning, Squawk Alfa 2506, continue* *climb 260.*

JP550 repeated the instructions: he would Squawk and climb to 26 000 feet.

Within seconds the code appeared by the relevant dot on the controller's radar screen, accompanied by the altitude of the aircraft and the flight number: and as it registered, the controller said:

10.03 38″	Zagreb	*that is correct, inbound Kostajnica, Zagreb,* *Graz next–*

For the crew of JP550, too, the familiar routine of yet another flight. At 10.04 hours they were 18 minutes airborne and levelling at 26 000 feet some 62 kilometres south of Kostajnica: airspeed 316 knots,[1] heading 359 degrees.

---

[1]The reader should understand that the airspeed of 316 knots is an Indicated Air Speed (IAS). At FL260 this would equate to a True Airspeed (TAS) (depending on air temperature and air density) of 460 knots (529 mph).

They would continue their climb as approved by ATC, until they reached their flight plan requested flight level of 31 000 feet on the Upper Blue Nine airway. By that time too, the Trident would have crossed the Yugoslav-Austrian border and joined the Upper Blue Five airway; height 33 000 feet, airspeed 295 knots (IAS)[2] and heading, 120 - 122 degrees, magnetic.

Both airways crossed over the Zagreb VOR beacon at Vrbovec.

JP550's Squawk code fell within the range of codes allotted to the middle sector and would not normally show up on the radar screens serving the upper and lower sectors. This was in accordance with 'Julia' – the computerised radar system used by the Zagreb Area Control Centre.

By this means, known as 'radar height filtering', the clarity of the radar presentation is preserved for each controller since he is able to choose the discrete airspace layer for which he requires a *total* display of information, namely, aircraft position, Squawk code identification and height. Aircraft operating in adjacent sectors of the Area (and therefore the responsibility of other controllers) would appear on his own screen merely as a blip without code or height read-out.

It is important to note this factor in the chain of events; it will consequently be useful to restate the upper and middle radar situation in the Zagreb Centre at that time:

On the upper sector controller's screen *all aircraft at flight level 310 and above were fully identified by Squawk code, position symbol – the 'blip' – and height.*
Aircraft operating *below* flight level 310 appeared on that screen simply as blips.
On the middle sector controller's screen *all aircraft operating at flight levels between 250 and 310 were fully identified by Squawk code, position symbol and height.*
All other aircraft appeared on that screen simply as blips.

It may also be noted that whenever necessary a controller may call up on his radar display the codes and height of aircraft outside his own sector. To do this, he must either feed the aircraft Squawk code transmitted for another sector into his own computer: ask the other sector concerned to request the aircraft to transmit the required code for his own sector: or use the 'Pointer' device which brings on to his radar display the position, height and code of a selected aircraft for a period of 30 seconds.

Shortly before ten o'clock Tann called Vienna, estimating Villach at 09.59. The crossword had by now been discarded: instead, the discussion concerned the merits of the Zagreb–Belgrade run.

'You ever done one of those Zagreb Belgrades?' Helm asked Tann.

'Oh yeah', came the reply.

[2]See preceding page. TAS here equals 480 knots at FL330 (552 mph).

'Bloody murder, aren't they', snorted Helm.

'Yeah', said Tann, indifferently. Plainly, he was not much exercised by the topic.

'You invariably get late, don't you? Some days you get the best part of your meal turnround there unless you want to get [back] a bit later.'

'I think most times, what I've done is to stop at Belgrade and perhaps have a meal, you know, first', said Tann. It wasn't that bad, his tone implied.

'You used to get quite a good meal there in fact', Tann added.

Helm sighed. 'Nevertheless, it's a bloody terrible day.'

'Yeah.'

'In the middle of a turnround. . .that sort of thing', pressed Helm.

He appeared to have touched a chord at last, for Tann voiced his own rueful complaint: 'Or you can spend I don't know what it is, about ten bloody hours on the aeroplane, literally', he said.

'That's right.'

'Yeah.'

They looked through the windows of the flight deck at the empty sky. There was silence for a while, broken by one or other of the pilots whistling and a few muttered comments: '. . .wants to go back in then?' and '. . .turbulence. . .'. There was the sound of switches. Then Acting First Officer Martin Flint (?)[3] said, 'Fuel's just about balanced.'

Tann gave a monosyllabic 'Yeah.'

They all laughed. Tann carried on. 'Yeah, Yeah, eh.'

But he had been listening to Vienna Control through his headphones for he said, 'Er, 1292, 476, thank you Wien, out'.

Someone coughed. It was 10.00 00″. Tann called Vienna again on the new frequency to report his position.

'Er, Wien, Bealine[4], gutentag, level 330, just turned over Villach, estimating Klagenfurt at 02', he said, then listened for the controller's reply. 'Roger', he said. He had been told to contact Zagreb upper sector on a frequency of 134.45 MHz.

'Er. Zagreb 134.45, thanks. Bye-bye.'

Tann had just told the Vienna air traffic controller that he was flying at flight level 330 and expected to be over Klagenfurt at 10.02. As instructed, he called Zagreb upper sector control.

10.04 12″	BE476	*Zagreb, Bealine 476, good-afternoon.*
	Zagreb	*Bealine 476, good-afternoon, go ahead.*
10.04 19″	BE476	*Er, 476, is Klagenfurt at 02, 330 and*
		estimating Zagreb at One Four.

[3]Voice not positively identified.
[4]Bealine: the call sign formerly used by British European Airways aircraft, usually followed by the flight number.

Chapter Seven

Something less than tranquillity now overshadowed the upper sector console, for the incoming announcement by BE476 had coincided with Mladen Hochberger's imminent departure from his post. His relief, Nenad Tepes, had failed to appear at ten o'clock and Hochberger had waited with a growing impatience which would not finally be denied. He would, he said, go and find Tepes.

Whatever else he said to Gradimir Tasic at this point is blurred by unresolved contest: but it is beyond dispute that from ten o'clock it was Tasic who had taken position as upper sector controller, and that from 10.05 and ten seconds it was Tasic – left without an assistant controller – who worked on, alone. And perforce, handled both jobs.

It was Tasic, therefore, who acknowledged BE476's entry into the upper sector at 10.04 12″ and it was Tasic who from that moment not only dealt unaided with the increasing number of aircraft in the sector but also maintained an intermittent liaison by telephone with Belgrade area control. It is instructive to examine the taped record of Tasic's work load at this point and to observe his application to the matters in hand.

	Zagreb	*Bealine 476, roger, call me passing*
		Zagreb, flight level 330, Squawk Alfa 2312
10.04 38″	BE476	*2312 is coming.*

Tasic had just confirmed that he had registered BE476's estimated time of arrival over Zagreb as being 10 One Four, and that it was flying at 330. The aircraft had now Squawked, and so was clearly identified on the upper sector's radar screen: next to the blip denoting the aircraft's position was the altitude, the flight number – BE476 – computed from the Squawk code, 2312.

Tasic was later to claim that the actual altitude shown on his screen was not 33 000 feet but '33 200 or 33 500' although both Munich and Vienna radar recorded 33 000 feet, thus confirming the pilot's read-out.

Nor, Tasic was to state, did he consider the discrepancy sufficiently important to warrant a query to the aircraft although regulations required such verification where the read-out differed by more than plus or minus 300 feet[1]

---

[1]Normal tolerance is plus or minus 200 feet.

from the level reported by the aircraft. For the time being, the radar was 'not dependable' and he would not use it.

But now there was little time to think about that. Within three to four seconds of the last call it became necessary to turn his attention to Turkey 889, which called in at 10.0441".

10.0441"	TK889	*Zagreb, Turkey 889[2] over Charlie 350*
	Zagreb	*Turkey 889, contact Vienna control 131*
		er . . . sorry, 129.2 Good-day.
10.0454"	TK889	*129.2. Good-day sir.*

Tasic's next recorded communication came 16 seconds later. This was a telephone conversation with Belgrade clearing Olympic 182 for the Sarajevo upper sector.

10.0510"	Belgrade	*Nasice, Upper . . .*
	Zagreb	*(I) need Sarajevo Upper*
	Belgrade	*Right away?*

Tasic was interrupted by Olympic's call:

| 10.0517" | OA182 | *Zagreb, Olympic 182, passing KOS at 05,* |
| | | *330, estimate Sarajevo 17.* |

To Belgrade, Tasic said: You can hear the message over the phone. . . . and to Olympic:

10.0520"	Zagreb	*Olympic 182, contact, Olympic 182 report*
		passing Sarajevo
10.0525"	Belgrade	*Hallo?*

Tasic showed some impatience. He had obviously been talking to the assistant controller: into the telephone he said:

10.0528"   Hallo, hallo, listen, give me the controller –

. . . but as the Belgrade controller answered, another aircraft called:

| 10.0530" | 9KACX | *Zagreb, Grumman 9KACX, with you,* |
| | | *flight level 410 –* |

Tasic, intent on clearing 182, did not immediately acknowledge. Instead, he continued his dialogue with Belgrade.

10.0535"	Zagreb	*Er, Lufthansa 360 and*
		Olympic 182–they've got nine
		minutes between them. Is that OK
		for you?

[2]As transcribed. Actual R/T call sign is Turkair.

45

Belgrade	*I've got it . . . OK*
Zagreb	*It's OK?*
Belgrade	*OK. It's OK.*

Tasic completed this call at 10.05 41" – in time to meet a new stream of reports from aircraft now in his sector.

10.05 44"	IR 777	*Zagreb, this is Iran Air triple seven . . .*
		good-after [sic] *. . . morning*
	Zagreb	*Go ahead, sir –*

He worked Iran Air 777 until the aircraft signed off at 10.06 12"; and dealt next with Monarch 148:

10.06 15"	OM 148[3]	*Zagreb, Monarch 148, we checked*
		Kostajnica 05, level 370, Sarajevo 19.
10.06 27"	Zagreb	*Monarch 148, report passing Sarajevo . . .*

148's acknowledgement gave the Grumman an opportunity to repeat his call.

10.06 37"	9KACX	*Zagreb, Grumman 9KACX is with you,*
		level 410.
	Zagreb	*9CX, good-day, maintain 410 and report*
		Delta Oscar Lima[4] and Squawk Alpha 2317

CX signed off at 10.06 59". The tape is silent now for 46 seconds.

What happened during that interval has yet to be described; but at 10.07 45" Tasic was back on the upper sector radio frequency:

| 10.07 45" | Zagreb | *Beatours 778, Squawk Alpha 2304* |
| 10.07 50" | BE 778 | *Alfa 2304 coming down, 778* |

Tasic would need to clear this traffic with Belgrade: again he spoke into the telephone. The tenor of this conversation is of special interest since Tasic – now clearly under pressure – was later to be accused of levity during this exchange. From the record then:

10.08 08"	Belgrade	*Hello?*
	Zagreb	*I need upper (clearance)*
	Belgrade	*Which one?*
	Zagreb	*Any one you offer me*

To Beatours 778 he said:

| 10.08 26" | Zagreb | *778 radar contact, continue* |

At the same moment, and as 778 acknowledged, he told Belgrade:

[3]Monarch.
[4]Dolsko.

46

| 10.0826″ | Zagreb | *You can hear in the receiver:* |
| | | *Beatours 778, radar contact, continue* |

So far all has been brisk and professional: but Tasic was now to encounter an apparently minor frustration which was subsequently to be unearthed and be most balefully interpreted.

| 10.0830″ | Belgrade | *Zagreb upper –* |
| | Zagreb | *Excellent! Estimating Bealine 476* |

A silence until:

| 10.0835″ | Belgrade | *Moment, please . . .* |

Tasic waited. It was a further ten seconds before he cleared his throat, as if to prompt the Belgrade controller.

| 10.0845″ | Zagreb | *Hm . . . hm . . .* |
| | Belgrade | *Nothing! I don't have anything (on BE476)* |

Tasic was sceptical: Nothing? he said.

| | Belgrade | *So nothing –* |

Shame on it! said Tasic: but at last the Belgrade controller, patiently checking his console, found what he was looking for.

| | Belgrade | *Here it is: Bravo Echo 476* |
| | Zagreb | *Yes* |

. . . and as Belgrade said: Further –? Tasic added:

		That's the Trident – yes: twenty-one from London to
		Constantinople
	Belgrade	*How do you spell Constantinople?*

Tasic wasted little time on his reply. Dryly he said:

| | | *Lima Tango Bravo Alfa [the I.C.A.O. designator* |
| | | *for Istanbul]* |

. . . which brought an equally dry repetition of those letters from Belgrade. On any other day that small jest might have gone unnoticed.

It was now 10.0903″.

From this moment, and until Tasic ended the call at 10.0955″, signing off with his phonetic initials 'Golf Tango', the controllers were continuously engaged in their task. Little more than 30 seconds lapsed before Tasic again returned to the upper sector radio frequency, at 10.1027″ to find himself about to deal with five aircraft (Beatours 778, Lufthansa 310, Finnair 1673, Olympic 172 and Beatours 932) in rapid succession. For this traffic, too, he would need clearance from Belgrade upper.

47

Tasic, it will be remarked, had been left alone at the console from 10.05 approximately. The pace of his activities since that time is clear.

He had cleared Turkair 889 for Vienna control, directed Olympic 182, telephoned Belgrade to clear both Olympic 182 and LH360, instructed IR777 to continue to Graz and to Squawk, told Monarch 148 to report when it passed Sarajevo and allocated Squawk codes to 9KACX and Beatours 778.

In addition, he had become involved in that lengthy telephone call to Belgrade, during which he had passed on the estimated arrival times for BE476 and 9KACX. He had been single-handed now for five minutes and 27 seconds: it was therefore, not surprising that when Erjavec asked Tasic to co-ordinate, the latter indicated that he was busy.

It had taken JP550 two minutes and 35 seconds to climb from 22 500 feet to the next ordered height of 26 000, and, on his arrival at this level Captain Krumpak reported in to the middle sector and indicated his readiness to continue the climb.

10.05 57″   JP550   *Adria 550, levelling 260, standing by*
                                              for higher

It was this message which was to bring together some of the apparently disconnected incidents of that morning and transfer them into a fatal loading of the scales: thus, it is on record that the middle sector controllers had been unable to plan in advance for the transit of JP550 since no flight progress slip had been prepared for the purpose. Erjavec's first intimation of the imminent entry of JP550 into the middle sector had come at the moment of co-ordination by the controller of the lower east sector, while the aircraft was in that airspace and crossing flight level 220. (It was that co-ordination which, despite the hiatus of the flight slip, was said 'not to have affected flight safety within the middle sector'. It was, however, to have a dire effect elsewhere.)

It is also on record that Hochberger had left the room in search of Tepes, and that he had found him about to make his entry; now – again contrary to regulations – the hand-over of the upper sector controller's traffic was being completed by these two men *outside the control room*[5] and without physical reference either to the flight slips on the console or the picture on the sector radar screen.

And Gradimir Tasic? Of course. At this time was he not both the air traffic controller and the assistant air traffic controller responsible for the upper sector of the Zagreb Area Control Centre?

While Tasic had been dealing with 9KACX, Erjavec, for his part, had found himself facing a problem.

JP550 had asked for permission to climb; but, as the slips on the console

[5]In fact, at the doorway.

confirmed, the flight levels above 260 were already occupied, 28000 feet by Adria 548 *en route* from Split to Nuremburg, and 31000 feet by Olympic 187, operating from Athens to Vienna. JP550, the controller decided, would have to climb into Tasic's sector.

Erjavec called the aircraft:

10.0603″	Zagreb	*550, sorry 330, er, 310 is not available, 3 – 280, also; are you able to climb, maybe to 350?*

Yes, said JP550 at 10.0611″, 'Affirmative, affirmative, with pleasure.' We can climb to 350.

Roger, said Zagreb at 10.0613″, Call you back.

'Yes, sir', replied JP550 at 10.0614″.

Raising his hand for attention, Erjavec looked over to Tasic.

The report produced by the Yugoslav Federal Civil Aviation Administration gives a brief account of what was said to have happened at that point. It notes that, according to the statement by the controller (Erjavec), the pattern of the co-ordination was as follows:

Erjavec signalled by hand to the upper sector controller, Tasic, indicating that he wanted to talk to him. Tasic waved his own hand in a gesture to show that he did not wish to be interrupted at that juncture since he was fully engaged with his traffic. Therefore Pelin, the middle sector assistant controller, went over to Tasic who was seated about 50 cm to his right in order to co-ordinate the climb of JP550 up to Flight Level 350. He asked Tasic for clearance for this, and at the same time showed him the aircraft by placing his own finger on the blip denoting JP550 on the radar display: after which, states Pelin, Tasic cleared JP550 for the climb.

These statements are not confirmed by Tasic who remembers only that Pelin came across to him and pointed to some aircraft near Kostajnica ...

It is only possible to fix the time of that co-ordination by reference to that 46-second gap in the record of Tasic's transmissions (from 10.0659″ to 10.0745″).

During that time, Erjavec had promised the pilot of JP550 that he would call him back, and at 10.0740″ he had done so, clearing the aircraft to climb higher. Presumably, therefore, Pelin had returned from his errand to the upper sector console to tell him that Tasic had now been briefed on JP550[6] and that he had been shown its position on the sector radar screen.

10.0740″		*Adria 550, recleared flight level 350 ...*
10.0745″	JP550	*Thank you, climbing 350, Adria 550 ...*

At 10.0750″, Zagreb telephoned Bec (Vienna) to alert the controller there of the decision.

---

[6]The extent of that briefing was later to be disputed.

Chapter Eight

2312 is coming, said Tann at 10.04 38″. Listening out on the upper sector frequency the Trident's crew heard Tasic working Turkair 889. Almost immediately they saw the other aircraft as it overflew them in the opposite direction.

(Cockpit Voice Recorder) There he is ...[1]

There was background conversation for a while.

At 10.07 31″, just at the time of JP550's clearance to climb to 350, Brian Helm made a morbidly prophetic remark while reading from a newspaper: 'A bit here, eight people injured in an aircrash on Tuesday were killed when a helicopter taking them to hospital crashed at ...', he said.

'Yeah, how unlucky can you get', agreed Tann.

'Obviously wasn't their week', Helm said. Then he changed the subject. 'Are you stocking up with vegetables for the winter?'

Stoically, Captain Tann accepted the new theme. 'Not really, one should I suppose', he said vaguely '... but –'

Helm spelled out the benefits: 'I was just looking at prices the other day of the first vegetables ... a pound in the market ... fresh ones down now about six or seven ...'

'Yeah', Tann said.

'So we're stocking up ... about the only things our kids will eat ... market in Farnborough just near us. You go along there about four o'clock, half past three or four o'clock, you get whole boxes of things ridiculously cheap ... a full box.'

'Yeah', said Tann. It was the barest encouragement.

'We bought a 13-pound box of tomatoes last week – 50p', Helm said, amazed. 'Yeah!'

'Most of them were green or yellow', Helm remembered.

'Yeah, yeah.'

---

[1]In the light of later comment, this tiny fragment is important: it shows conclusively that the British pilots were doing what they were meant to do, i.e. monitoring the R/T, listening out and looking out.

Chapter Nine

At 10.09 18″ the middle sector controller glanced at the radar screen and identified JP550 by its Squawk code, Alpha 2506.

He noted that the aircraft was nearing Kostajnica.

Zagreb		*550 approaching Kos, proceed to Zagreb, Graz, and call me passing 290 –*
JP550		*Roger*

and at

| 10.0949″ | JP550 | *Zagreb, Adria 550 is out of 290 –* |

The next step in the climb would bring 550 into the upper sector at 31 000 feet. Erjavec asked for a final call –

| 10.0953″ | Zagreb | *Roger, call me passing 310 now –* |
| 10.0955″ | JP550 | *Roger* |

It took two minutes and 48 seconds for JP550 to climb from level 290 to 310.

10.1203″	JP550	*Zagreb, Adria 550, out of 310.*
10.1206″	Zagreb	*550, for further Zagreb 134.45. Squawk stand by and good-day sir.*
10.1212″	JP550	*Squawk stand by, 134.45. Good-day.*

In this exchange Erjavec had instructed JP550 to switch to the upper sector frequency of 134.45 MHz and to Squawk standby: in other words, to switch off the transponder. He had done this since the Squawk code, it will be remembered, was specific to the sector. The middle sector had been allocated the range of codes from A2500 to A2577 and the upper sector, the codes from A2300 to A2377.

In telling JP550 to switch off the transponder, therefore, Erjavec had done no more than release his own code during the transfer, and – since this procedure was not regulated by national or international requirements – had acted within a proper discretion.

As JP550 complied, the symbols denoting its code and altitude were deleted from the middle sector radar display and only the blip indicating the aircraft's position remained.

JP550, of course, was also merely a blip on the upper sector radar: that blip which had been 'fingered' by Pelin during the co-ordination.

There had been little else to prepare Tasic for the coming of JP550, for if, at last, he received a flight slip for the aircraft – possibly at 10.1212″ as JP550 changed to the upper sector frequency – he had scant time in which to consider it: for at 10.1210″ he was again fully engaged in a concentrated exchange of information.

| 10.1210″ | Zagreb | *Finnair 1673, report passing Delta Oscar Lima,*[1] *maintain level 390, Squawk Alpha 2310 –* |

An eight-second gap. Then:

10.1220″	F1673[2]	*Will report passing Dolsko at 390*
10.1224″	LH310	*LH310, Sarajevo at 09, 330 Kumanovo, 31*
	Zagreb	*LH310, contact Beograd, 134.45 – sorry, – sorry, 133.45 – good-day.*
	LH310	*Good-day.*

Good-day, repeated Tasic at 10.1238″. And now it was Olympic's turn.

| 10.1240″ | OA172 | *Zagreb, Olympic 172, good-afternoon, level 330 –* |

Another silence: possibly because Nenad Tepes had now joined him at the console.

10.1248″	Zagreb	*Olympic 172, go ahead*
	OA172	*Olympic 172, level 330, estimate Dolsko at 16*
10.1300″	Zagreb	*Olympic 172, report passing Dolsko, flight level 330, Squawk Alpha 2303 –*

10.1307″ Olympic repeated the message and confirmed 'Dolsko direct to Kostajnica' –

10.1315″	Zagreb	*Affirmative sir*
10.1318″	OA172	*Thank you*
10.1319″	BE932	*Zagreb, Beatours 932 is level 370, estimate Dolsko 18.*

Again a silence – perhaps because Tasic would need to familiarise Tepes with the situation in the sector?
932 called again.

| 10.1334″ | BE932 | *Zagreb, Beatours 932 –* |
| | Zagreb | *962, go ahead.* |

Thus, Tasic, increasingly harassed, made his second verbal slip in just over one minute.

[1]Dolsko.
[2]The Finnair flight prefix is AY: the call sign is Finnair. The letter 'F' above is as transcribed.

52

| 10.13 42″ | BE932 | *Beatours, 370, estimate Dolsko 18* |

Nenad Tepes had now begun the task of clearing the upper sector traffic with the Belgrade control. There is a break of 11 seconds on the upper sector radio frequency tape before Tasic's voice is heard once more:

| 10.13 53″ | Zagreb | *Beatours, maintain flight level 370 and report overhead Dolsko. Squawk Alpha 2332 –* |
| 10.14 03″ | BE932 | *Roger, 2332* |

In JP550, Captain Krumpak had switched to the upper sector frequency as ordered: but owing to the continuous flow of R/T traffic between Tasic and other aircraft, he had been unable to make contact with the upper sector controller. Meanwhile he had continued his climb until, now some 250 feet below the approved level of 350, he at last managed to raise Tasic:

10.14 04″	JP550	*Dobar dan Zagreb, Adria 550*
		(Good-morning Zagreb, Adria 550)
10.14 07″	Zagreb	*Adria 550, Zagreb dobar dan (good-morning)*
		Go ahead.
10.14 10″	JP550	*325 crossing, Zagreb at One Four*

On the upper sector radar screen the single point of light that was JP550 now hung like a satellite beside the Squawk symbols of an aircraft approaching from the opposite direction.

BE476   Flight level 330, estimate Zagreb One Four

With sickening clarity an awful realisation flooded Tasic's mind. Urgently, he called:

| 10.14 14″ | Zagreb | *What is your present level?* |

... and heard Krumpak's matter-of-fact reply –

| 10.14 17″ | JP550 | *327* |

As a man in a nightmare, Tasic saw what must now happen, and made a despairing attempt to avert it.

He gripped the microphone and, as the unwitting Tepes beside him continued his grotesquely placid accompaniment, said rapidly:

| 10.14 22″ | Zagreb | *[stammering] ... e ... zadrzite se za sada na toj visini i javite prolazak Zagreba (... e ... hold yourself at that height and report passing Zagreb).* |

In his agitation his English had deserted him: he had spoken in Serbo-Croat.

Krumpak, puzzled, replied in the same language:

| 10.14 27″ | JP550 | *Kojoj visini? (What height?)* |

Stay where you are! prayed Tasic silently. Please stay where you are. Eyes riveted to the radar screen he watched, helpless, as JP550's blip crept into the corona of the other signal: and sought, agonised, for words:

| 10.14 29″ | Zagreb | *Na kojoj ste sada u penjanju jer . . . e . . . imate avion pred vama na isn . . . [voice not coherent] 335 sa leva na desno. (The height you are climbing through because . . . you have an aircraft in front of you at . . . [voice not coherent] 335 from left to right.* |

Olympic 172, Tepes was saying, 330. Next Beatours 932.

| 10.14 38″ | JP550 | *Ok, ostajemo tocno 330. (OK, we'll remain precisely at 330 –).* |

And, as Tasic watched, both targets merged and became a minute single flare which brightened momentarily, became again two points of light, then died.
Sick at heart he began to call

Bealine 476, Zagreb, report passing Nasice

Mechanically he fended off calls from BE778 and Iranair triple seven; and returned again to the empty silence.

10.15 50″	Zagreb	*Adria 550, Zagreb*
10.16 00″		*Adria 550, Zagreb*
10.16 14″		*Adria 550, Zagreb*
10.16 32″		*Adria 550, Zagreb*
10.16 42″		*Adria 550 Zagreb*
10.16 50″		*Bealine 476, Zagreb*
10.16 58″		*Bealine 476, Zagreb*

Chapter Ten

The pilot of Lufthansa 360 saw it happen. Flying towards the Zagreb beacon at 29 000 feet on Upper Blue Five, some 15 miles behind the Trident, Captain Joe Kroese glimpsed the sudden flash, as of lightning, and the ball of smoke from which emerged two plunging shapes.

It was his excited call which broke in on the middle sector frequency while on the adjacent console Tasic strove with his insistent and hopeless search.

10.15 40″   LH360   *... e Zagreb! It is possible we have a mid-air collision in sight – we have two aircraft going down, well, almost below our position now –*

The words failed to register. Frowning, the middle sector controller said:

10.15 52″   Zagreb   *Yes, two aircraft are below you, but I don't understand you: what do you want, sir?*

Almost frantic, Kroese repeated his message:

I think there's been a mid-air collision! – Two aircraft are going down with a very fast rate of descent – it might be a fighter but I think it might also be an airliner –!

I'm sorry sir, said the controller. I don't understand you.

Yet again Kroese attempted to report what he had seen: and yet again confusion reigned on the frequency as the controller became increasingly baffled.

10.17 19″   LH360   *Do you still have contact with Olympic airliner?* asked Kroese
Olympic was on course for Graz, he was assured.

Not him! said Kroese: the aircraft ahead of us! I believe it's 172 –?

It was not until 10.18 02″ that Captain Kroese was asked to repeat his message for the last time.

Lufthansa 360, this is Zagreb: will you be so kind and say again? Do you have a problem?

10.18 12″	LH360	*We don't have any problem,* said Kroese: *But in front of us about 15 miles or so I think we did see a mid-air collision. It's possible that the other aircraft ahead of us had a mid-air collision ... er ... just overhead Zagreb. We had two aircraft going down with a rapid rate of descent ... and there was also smoke coming out.*

As the words sank home at last the middle sector controller looked to his right. He was in time to see Gradimir Tasic, dazed and ashen faced, in the act of raising both hands to his earphones and, very slowly, laying the headset down on the console before him.

Chapter Eleven

JP550 and BE476 collided above the Zagreb VOR at a combined speed of more than eleven hundred miles per hour. In that tremendous impact some five metres of the DC9's port wing cut through the base of the forward windows of the Trident's cockpit and continued on like a gigantic scythe through the flight deck and forward passenger compartment. Hurtling debris smashed into the Trident's rudder and carried it away, while the hull itself, now shorn of the forward section of its cockpit, began a steep dive to the ground.

The DC9, its port wing completely severed, burst into flames. Torn metal penetrated the port engine which, in turn, flung the shattered remains of its compressor blades into the empennage. This section, with its disintegrated tail surfaces and cone portion containing the aft staircase, broke away as the aircraft fell.

With it fell a ghastly rain. Those who witnessed it would remember it for the rest of their lives.

Walking home near the cornfields of the village of Gaj, near Vrbovec 30 kilometres north-east of Zagreb, 21-year-old Luba Panketic 'heard a tremendous noise . . . like a thunderclap. Then the body of a young girl crashed to the ground five yards in front of me. I screamed in horror. Next, suitcases started falling around me and I began to run. I almost fell over the body of a man which fell from the skies.'[1] Other terrible things plunged into the ground. In seconds, the earth became a bloodied nightmare of strewn bodies strapped into aircraft seats and the unspeakable, obscene fragments of mutilated humanity: for now, before this girl's terrified gaze lay 'wreckage . . . and human arms, legs and heads everywhere'.[2]

Farmer Stjepan Cicek looked up as his 12-year-old son cried out: 'Dad, Dad, bombs are falling!' He was in time to see a man falling out of the sky.[3] 'First he lost his jacket, then the shirt', said Cicek, wonderingly, and thereby revealed the depth of that frightful imprint on his mind.

Three women in Gaj saw a little girl plummeting down: she lay where she had fallen and 'we covered the little one up, because she no longer had any clothes on'. They were told to let her lie there: 'so . . . we laid some flowers down

[1]Reported in *Sun* 11 September 1976.
[2]*Ibid.*
[3]Reported in *Kolner Stadt-Anzeiger* 13 September 1976.

too, and watched all night'.[4]

In the nearby village of Luka, two women heard something like a cannon-shot in the air and saw the merged vapour trails spill a dreadful cascade into a field just a kilometre from where they stood. With a group of farm workers they ran to the spot 'to see if they could help: but it was too late for that: only a child of five, strapped in its seat, still showed some signs of life . . . but just afterwards, it died'.[5]

Others too, had heard that massive thunderclap, had likened the moment to 'a gigantic bang, then a fireball from which hundreds and thousands of pieces fell to the ground'.[6]

In Vrbovec, technician Zivko Koretic stood shocked and deafened by the explosion: then he looked up and saw 'two planes rocking and floating in the air. I knew at that moment they were in collision. One plane – I realised later it was the Yugoslav DC9 – burst into flames and went down sharply. The Trident kept afloat for some time, as if the pilot was trying to save it.'[7]

But there was nothing to save.

'Suddenly I heard a terrible crash', said a farmer in Luka. 'And then I saw an aeroplane burning like a torch and falling apart as it came down.' It was the Mayor of Gaj, however, who also 'heard a crash like thunder and rushed out to see the wings of one plane tumble out of the sky', who recovered himself sufficiently to sound the emergency alarm within three minutes of that first explosion. About 15 minutes later the first rescue vehicles arrived, sirens wailing, at the remote crash sites. Among their crews was a doctor from the Drugo Selo medical service, who recounted bitterly: 'I couldn't help anyone. There were just no signs of life.'[8]

The two machines came down within seven kilometres of each other.

The Trident had crashed tail first and starboard wing down. It lay now amid the hideous litter of the cornfield, some one and a half kilometres to the south of Gaj. Smoke, soon to leave only a pungency in the air, drifted from the silent power plant: no other movement came from what had been BE476.

The burning DC9 ploughed into a forest one kilometre east of the village of Dvoriste, igniting about seventy square metres of the area and scattering disintegrated parts of the aircraft for a distance of two and a half kilometres. The main body of the fuselage, grotesquely inverted, burned out completely.

The 'Preska' group of twelve volunteer firemen who arrived at the crash site eight minutes later did what they could with water. It was not, however, until they had been joined by a regular and fully equipped fire brigade summoned

[4]Reported in *Kolner Stadt-Anzeiger* 13 September 1976.
[5]*Ibid.*
[6]*Daily Express, Daily Mail* 11 September 1976.
[7]Reported in *Kolner Stadt-Anzeiger* 13 September 1976.
[8]*Ibid.*

from Zagreb that the blaze was finally brought under control.

Soon, by 11.15 there was only the dead and steaming ruin of JP550 lying in its blackened swath.

When there was nothing more to be done at the crash sites, men began a wider search for survivors. At both places small groups of firemen, ambulance crews and local people had combed the immediate vicinity but now the number of those involved grew, until, as they moved slowly over the rain-wet terrain the organised deployments of police, medical teams, soldiers and villagers totalled twelve hundred persons.

There could be little hope of finding anyone alive, yet the cornfield and the surrounding area was probed for two full days and the forest for three before the pitiful harvesting was abandoned.

It had indeed been a pitiful task: of 176 people only two had lived, briefly, beyond the final impact: yet there was much else for them to find. The search parties picked their way over fields and ploughland and among the splintered timber and gathered what they could of the bodies: mercifully, thereafter, they could focus their attention on the gathering of other things such as clothes and toys and magazines and sundered luggage, much of this – suitcases and bags – retrieved from rooftops and farmyards. It should have been enough: but on *his* roof, farmer Stjepan Mojdak found a dead woman still strapped in her seat and in his neighbour's garden, more bodies; and in a barn in Gaj, a torn-out aircraft seat, also with its occupant, which had crashed through the roof.[9]

The full horror was thrust upon the appalled British Consul in Zagreb, David Montgomery, who reached the scene soon after the crash: '. . . there was a small girl, between six and ten. I saw her . . . senior police officials told me that she was still alive about an hour after the crash. There was another woman still in her seat belt who survived for about half an hour. I think these were the only people living after the crash.' Many of the bodies of the passengers were still inside the aircraft: 'Fifty per cent were still inside, 25 per cent were scattered near the wreckage and the remainder about 200 yards away.'[10]

Parts of the aircraft together with bodies and parts of passenger bodies were found in the area between Krkac/Graberanec to the north, and Pirakovec/Vrbovec to the south. Passenger baggage and pieces from both aircraft were also found, while light material was scattered over the area that extends from north-east of Vrbovec to eight kilometres north-east from Krizevci – a distance of 32 kilometres.[11]

Later, a British journalist visited the sites. Of the DC9, almost nothing was recognisable: but

On the side of the narrow road leading to the village of Gaj, where the

[9] Reported in *Kolner Stadt-Anzeiger* 13 September 1976.
[10] *Daily Mail* 11 September 1976.
[11] Yugoslav Federal Aviation Administration – Aircraft Accident Investigation Commission (Kromisija za Ispitivanje Avionskih Nesreća Savezne Uprave za Civilno Zrakoplovstvo).

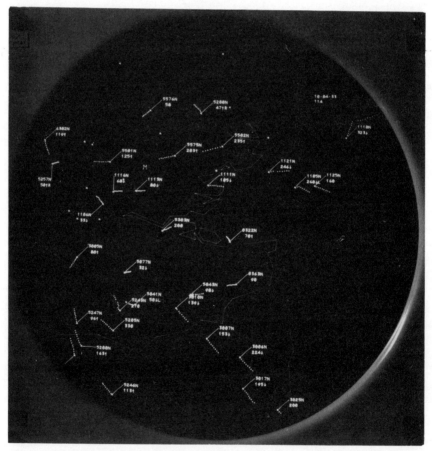

Plate 1 Secondary radar presentation showing aircraft targets with individual identification and height.

Plate 2 British Airways Trident Three.

Plate 3 Inex Adria DC9.

Plates 4, 5 Wreckage at crash site.

Trident crashed down in maize and pumpkin fields, are items from the aircraft's galley. Debris lies scattered amongst the acres of six-foot-high maize plantations. The cockpit is some 300 yards ahead of the plane's bulk. The control columns are smashed and the multitude of instruments shattered. A stewardess' blue shoe is caught among the jumble of torn fabric, power lines and equipment. Nearby are yellow lifejackets, flight documents and more personal effects.

It was a succinct and moving glimpse of a terrible destruction, left now to the army and civilian rescue teams. Stolidly they worked on; and on again under the glare of searchlights as the day darkened. Rain slanted through the beams as they hauled and shifted through the night.

Now too, as telephones and teleprinters began to spread news of the event throughout Europe and beyond, came the questions. And inevitably, these were posed first by those who dealt, also, in assumptions.

In this category were the hungry newspaper and television reporters who required immediate answers and identifiable causes. They pressed, therefore, upon a second group of questioners – the executives of the respective airlines, airport officials and officers of the Accident Investigation Branch of the United Kingdom's Department of Trade and the Yugoslav Federal Aviation Commission. These men, too, required answers and in the course of time would provide them. But this would wait until they had carried out their primary duty of establishing the facts; and until that time, the Press – although there were notable and honourable exceptions – could, and did, pronounce unchallenged.

Another group, too, was concerned with fact although one of its leading members was unwary enough to offer what appeared to be a prior opinion to the eager reporters.[12] These men were representatives of the Yugoslav judiciary: for the five air traffic controllers of the morning shift at Zagreb Centre had immediately been 'detained' and, within hours, had been formally declared under arrest.

[12]*Daily Telegraph* 13 September 1976: see also pages 67-8.

Chapter Twelve

In the control centre, work came to a standstill as the import of the message from Captain Kroese was fully comprehended.

For Julije Dajcic, the shock of the news brought physical collapse; for one of his men, an incoherent paralysis; for the rest of the day shift, the demoralisation of self-doubt and fear; and for the female staff – the Flight Data girls and others – the undeniable and perhaps more merciful response of tears.

Yet, within a short time, firm decisions emerged from the chaos. The full shift were relieved of their duties and warned to await the arrival of the police; to replace them, all off-duty controllers were summoned and transport was provided to round them up.

One hour later, the consoles were newly manned and Zagreb Control was once more operational. The men of the original shift however remained in limbo, each of them preoccupied by the implications of his own situation: Dajcic, responsible as supervisor; Tepes who had been late; Hochberger, who had left the console; Pelin who had handed JP550 over to Tasic; Erjavec who had told him to do that; and Tasic, for what he had done or failed to do.

Now they had undergone their first interrogation at the hands of the police. They would, they were informed, be held for further questioning.

The news came less swiftly to those who waited for JP550 at Cologne/Bonn Airport.

> ... those who were waiting for the return of relatives ... and those who had originally planned to fly to Yugoslavia on board the plane which had now crashed, were only told of a delay 'due to technical faults'. The flight, which was to have arrived at 12.10 p.m., had been 'forced to land at Zagreb because of a defect' announced the loudspeakers shortly after noon. It was now expected to arrive at 17.10. Many people from Cologne decided not to wait at the airport until 5 p.m. and drove back to the city: but others felt that it was not worth the trip and sat patiently in the airport restaurant for which they had been given refreshment vouchers.[1]

And meanwhile the authors of those announcements phoned frantically for news, continuing until the major details had been established. Dispiritedly

[1]Reported in *Kolner Stadt-Anzeiger* 11/12 September 1976.

then, they began to put together their next announcement.

Some would find the truth in other ways. Shortly after 3 p.m. a housewife from Koblenz who had come by car to meet her nephew went out to put some more money in the parking meter because of the delay. On the way she told one of the car-park attendants that she was waiting for the plane from Yugoslavia and, until he spoke, failed to realise the significance of the tiny transistor radio beside him. 'But that's crashed – haven't they told you?' he asked in surprise.[2]

But at the travel company's desk, they knew nothing; only that there had been a delay ... There was nothing more for her, and for what was now a growing number of enquirers, until, at 4.30, a duplicated sheet was handed out at the desk.

Its message was brief. It stated that a DC9 belonging to the Yugoslav airline Inex-Adria had crashed after a collision which occurred after the plane 'had reached a cruising height of 10000 metres near Vrbovec in the Zagreb area'. '... the woman went to the telephone booth with her sister and a brother who had arrived by then, in order to tell her own children the news. She broke down at the telephone and had to be taken away on a stretcher.'[3]

The announcement of the delay had not, at first, worried young Rolf Hanneman, of Bonn, who had come to meet his parents. He had sat at the bar and then gone for a walk; and had returned to find the crowd at the desk. 'Then he also saw the sheet which people were holding and which carried the news of the crash. He turned away and went towards the exit. "I don't know", he mumbled. "I'll have to see –". He struggled to understand that his parents, who had so often gone on holiday by plane and always returned safe and sound, would not be returning this time.'[4]

By now, as the news spread through the airport, there were others who shared a similar knowledge. Some, like Hanneman, stumbled blindly away to grapple alone, or at least elsewhere, with the awful fact of their loss. Others remained as if there could be anything more to learn or to revoke what they had heard. Among them were couples and numbed family groups who could hold back nothing of their grief and tears.

The news reached British Airways in London at 1.30 p.m., two and a quarter hours after the event, and some 45 minutes after the Trident had been due to land at Istanbul.

They would question the reasons for that belated intelligence, said the airline, but wasted no time in bringing their own emergency procedures into operation. Accident information centres were set up in London, Istanbul and Zagreb: at Istanbul Airport, Alex Workman, District Superintendent, began dealing with enquiries and tracing next-of-kin while David McMillan,

[2]*Ibid.*
[3]*Ibid.*
[4]*Ibid.*

Manager of Austria, Hungary and Yugoslavia divisions, abandoned his holiday to go to the Zagreb accident centre.

A special telephone number, which people could ring for further information, was broadcast on television and radio and appeared in the newspapers. A 24-hour shift system was organised and personnel staff, drawn from London Stations management, and the passenger-handling and administration sides of Terminals 1 and 2, were mobilised. Almost at once, the flood of incoming telephone calls began as anxious relatives and friends commenced to make enquiries. Arrangements were made by airline staff to meet passengers flying in, to find hotel accommodation for relatives and to advise on funeral arrangements.

Roger Phipps, Manager of Terminal 1 at Heathrow, explained how the system worked. 'A list of names of passengers and crew has to be obtained, and cards drawn up for each person. These are constantly up-dated as information comes in and the next of kin are informed . . . We received an enormous amount of support from staff offering help – many came in from their homes.'[5]

At Heathrow, too, the bitter scenes of Cologne were re-enacted. Flight staff were shocked and many of the stewardesses cried: some were led away by colleagues. One of the girls said: 'In this job you are in constant fear of something like this happening. When it does happen, you don't thank God that you weren't on the flight. You think about friends and pray that that they are all right. You are always involved.' Another added, 'You often think about an accident, but when it does occur it seems to be unreal. When it sank in I just broke down and cried. All the girls are shattered.'[6]

It was no less painful for Sir Frank McFadzean, Chairman of British Airways, for Managing Director Henry Marking and for the Chief Executive of the Airlines European Divisiosn, Roy Watts. After an urgent meeting at the Heathrow Centre, together with management representatives from Operations Control and Flight Operations, Watts left on a specially organised flight for Zagreb. With him went a six-man team which included Air Safety, Flight Operations staff and a doctor. They were joined by four members of the Accident Investigation Branch of the U.K. Department of Trade, headed by Mr Colin Allen.

Both teams were to work with the Yugoslav authorities in a common endeavour to find out what had actually happened.

[5]*British Airways News* 17 September.
[6]Reported in *Daily Express* 11 September 1976.

Chapter Thirteen

By 10.30 p.m. on 10 September, most relatives had been informed of the crash. The passenger list was not announced in full until this task, greatly complicated by the mix of nationalities represented, had been completed. By that time too, the accident had been headlined in the evening newspapers. Based on whatever could be gleaned from any source, the early reports deserved credit, as always, for the comprehensive nature of the information presented. As always, too, however, and as if in fear of any admission of less than total infallibility, any potential hiatus in the flow of fact was signalled by the presence of some desperate and unfortunate guesswork. Thus, the banner headline of the *Evening News* of that date proclaiming '171 killed in worst collision' was contradicted by its companion newspaper the *Evening Standard*, with the correct figure: 'Mid-air crash – 176 die . . .'; and while both newspapers offered impressive coverage of the collision and its location, the convening of the British investigation team, the views of meteorological officers and whatever else of aviation and human interest concerning the aircrew and passengers could be obtained, both featured the implication that would later crystallise so brutally into the indictment of those at Zagreb.

Interviewing 'a technical expert who knows Zagreb Airport', the *Evening Standard* reported him as saying: 'Zagreb is a cross-roads and presents a case of potential air congestion. Near the city, three trunk routes cross over at a beacon – it is exactly like the cross-roads in a town, although I cannot say whether this had anything to do with the collision.'

More bluntly, the *Evening News* posed the word 'Error?' in a bold cross-heading, and below it, stated: 'One possible cause could be an air traffic control error, but no evidence has yet been revealed.'

It is more than possible that these statements owed their origin to an apparently authoritative, and to some surprised ears certainly premature, observation by Judge Vjeceslav Jakovac, appointed by the Yugoslav authorities on that day to carry out a preliminary investigation into the accident. Judge Jakovac had personally ordered the detention of the five air traffic controllers after hearing tape recordings of their conversations during the shift: that evening he had appeared on television.

Confirming that 'some people' had been detained for questioning, he added: 'It probably happened because of incorrect assessment of the altitudes of air corridors for the two planes.'[1] In the eyes of the public this remark pointed to, if it did not entirely determine, the orientation of the subsequent investigation and, indeed, was reinforced a day later when the Judge visited the crash site. 'This is a very serious matter', he commented, 'when you find men who are directly responsible for a disaster of this magnitude . . . Having seen the awful, gruesome debris of the two airliners, it gives me an even heavier heart to admit this.'[2]

It was clear that Judge Jakovac had already made his own decision on what should be the major area of investigation and towards whom any forthcoming blame should be directed. Further emphasis appeared to be unnecessary: yet, it was nevertheless supplied: 'Standing by the burned-out wreckage, the District Judge of Zagreb, Vjeceslav Jakovac said: "I cannot say definitely and officially whether it was pilot error or control tower error. But most probably, the cause of the disaster was a misjudgement of altitude and timing in the air corridor." '[3]

Given these leads it seemed irrelevant to discuss other possibilities and early speculation on the theme of 'pilot error', 'mechanical defect' or 'congestion of the air corridors' disappeared from the columns of the newspapers thereafter.

The omission however, was coupled with another, and more serious, default: for none of the British newspapers saw fit to suggest that the words uttered by Judge Jakovac might also have prejudiced an objective analysis of the cause of the accident and contributed in no small manner towards the damnation of men as yet untried.

That issue was left to the Yugoslav newspapers themselves to ventilate, and it is ironic that they did so only after official discouragement of their own attempts to reproduce sketches of the possible sequence of events which had already been published by the Western Press.[4] Their protest came after the Yugoslav Government had, for reasons of its own, 'urgently recommended' them to discontinue the publication of such matter and was couched in no uncertain terms. The advice, they retorted, was a 'muzzle'. 'It was a thankless task to accuse only the air traffic controllers under arrest, since they were doubtless just "partly to blame" and . . . unimportant cogs within the great machinery of responsibility.'[5]

It is inescapable that whatever was to be the outcome of a final enquiry, the controllers were adjudged guilty on the day of the crash.

The men had been under continuous interrogation since noon on Friday 10th and would now, on the 13th, appear before a tribunal headed by Judge Branko Sarav. 'The tribunal will prepare a report on the evidence for the public prosecutor. If he decides that they have a case to answer on charges connected with endangering public safety and causing death by negligence, they will be

---

[1]Reported in *Daily Mail* 11 September 1976.
[2]*Daily Mail* 13 September 1976.
[3]The *Sunday Times* 12 September 1976.
[4]See also page 123.
[5]*Der Spiegel* Nr 22/1977.

tried in public and could face prison sentences ranging from three to twenty years.'[6]

That possibility did nothing to inhibit correspondents and those others designated merely as 'spokesmen', nor did the qualification in the same statement that it would be necessary to consider the evidence. On the contrary, the published comments of Judge Jakovac were eagerly seized by the Press and within a very few days – fed by every new scrap of 'information' and conjecture – were developed in such fashion as to exclude any kind of reservation.

It is clear that a climate of opinion was now being generated of which the chief ingredients were the moral destruction of the controllers and a widespread acceptance of their culpability: and it is equally clear that this owed much to inferences too quickly drawn from such comments. The fact that these were properly based on an inquisitorial system of investigation quite different to the legal process of Britain was, in general, obscured by more dramatic and emotionally stimulating material.

With application, it was still possible for more discerning readers to claim impartiality for the Judge's announcement that the detained men were being held, temporarily 'on the basis of justified [sic] suspicion that they were responsible for this disastrous accident'.[7] But those who yet harboured doubts could refer to another and, for the air traffic controllers, similarly menacing translation: 'The Judge said that they were being held on justified suspicion that they are personally [sic] responsible for the disaster.'[8]

It is unfortunate that protests by the International Federation of Air Traffic Controllers' Associations concerning the legitimate interests of their members impinged, for the most part, on other members of that profession. For the rest, the path had been signposted and, little more than one week after the collision, journalists were able to tread, sure-footed at last, out of the maze of possibilities. The way had been made even easier by the speedy decision of the examining Judges that there was, indeed, a case to answer, and that the five controllers would therefore be sent for trial before the Zagreb District Court.

It would be many months before that could happen and by then, only one of the accused, Tasic, was still in custody:[9] but by that time, too, the public, in order of priority, had not only already heard the verdict; they had also been informed of the 'reasons' for the crash. It did nothing to diminish their authority or even seem important that not a single word of evidence had yet been published to support either of these assertions.

While other British newspapers were still reporting that 'legal action began yesterday' the *Daily Express* of 14 September featured the result of that

[6]*Daily Telegraph* 13 September 1976.
[7]Reported in the *Observer* 12 September 1976.
[8]*News of the World* 12 September 1976.
[9]Dajcic had been released within a day or so of the crash. Tepes, Hochberger and Pelin were released on 11 November since the legal limit of two months custody had expired and the Public Prosecutor had not requested its extension. Reported in the *Guardian* 12 November 1976.

investigation in a short paragraph: 'The Yugoslav judge heading the inquiry into the mid-air collision between a British Airways Trident and a DC9 tonight put the blame squarely on the shoulders of four of the arrested five control tower staff at Zagreb.'

The report contrasted somewhat with a paragraph in *The Times* of the same date which stated merely that: 'Mr Vjeceslav Jakovac, the examining magistrate, disclosed at a press conference that four of the five had been detained. Mr Jakovac said they had been held because there were grounds for believing they were in contact with the crews of the British Airways Trident and the Inex-Adria DC9 when the accident occurred.'

There were fewer concessions to prudence on the 15th however, for the reiteration of the verdict was now accompanied by a considerable element of hypothesis.

The mid-air collision over Yugoslavia last week which killed 176 people is understood to have resulted from a simple mistake by a Yugoslav air traffic controller which was compounded by a breach of air traffic regulations.

International rules state that all conversations between aircraft pilots and control tower staff should be in English. But, according to a West German source, Zagreb control tower staff[10] were using Serbo-Croat to to pass instructions to the Yugoslav DC9 involved in the collision.

If English had been used, the crew of the British Airways Trident ... might have realised that a mistaken instruction had been given to the DC9 and taken action.

The original mistake thought to have been made by Zagreb control was clearing the DC9 to fly right into the path of the Trident.[11]

The theory of language confusion as the cause of accident was even more strongly presented elsewhere: yet while it remained a reasonable possibility, it was nevertheless stated as established fact.

A radio message could have prevented the Yugoslav mid-air disaster in which 176 people died. But the pilot of the British Trident couldn't have understood it – it was spoken in Serbo-Croat.

The message – directing the pilot of the Yugoslav aircraft on to the course that brought it into collision with the Trident – should have been in English.

But air traffic controllers broke this local rule, it was claimed today. As a result, Captain Dennis Tann never realised he was on a collision course last Friday. The British captain would almost certainly have been listening in to the instructions given to the DC9's pilot, Joze Krumpan [sic] – but he did not know the language.[12]

[10]Note plural.
[11]*Daily Telegraph* 15 September 1976.
[12]*Daily Express* 15 September 1976.

In both newspapers the suggestion was given extra weight by the revelation that 'Captain Joe Kroese, the pilot of a Lufthansa Jet which was below the Trident and the DC9 jets at the time ... has reported overhearing radio instructions being given to the Yugoslav crew in Serbo-Croat.'

Little had been left to the imagination: but if the point required any further emphasis it was unwittingly provided by the inclusion of a quote by Captain Peter Harper, a Trident pilot and the British Airline Pilots Association representative at the preliminary enquiry in Yugoslavia. 'English is the recognised international language but I have heard air traffic control staff talking in French and indeed, in Serbo-Croat.' He added: 'It's wrong, but pilots tend to let it go as long as it doesn't affect them.'[13]

By the end of that week the event had been encapsulated in a common sentence shared by a number of British newspapers: 'Yugoslav air traffic controllers cleared both planes to fly on a collision course.'[14]

It served as an explanation. That it anticipated both the official report which was published one year later in October 1977 and the outcome of the legal ordeal which the controllers were yet to face, is a matter of record.

But, by 19/20 September, too, more specific information had been released by the crash investigators who had now, themselves, disposed of other possibilities:

Investigators probing the world's worst mid-air collision ... now believe the local air traffic control 'lost' the British airliner involved.

This they think, is the most likely explanation of why the other aircraft ... was ordered to fly right through the path of the ... Trident ... Much of the wreckage has been examined and the investigators, including a British team from the Department of Trade, have found no signs of anything wrong with either aircraft before the collision. Everything points to a traffic control error.[15]

That statement was, of course, a remarkably accurate visualisation of what had actually occurred. It did not, and could not, tell the full story since this was yet to be unravelled: but the investigators were confident that after ten days of patient and sometimes frustrating enquiry they had at last narrowed the field.

It had been difficult, notwithstanding the courtesy and co-operation of their Yugoslav hosts since it had also been necessary for them to thread their way through diplomatic niceties, preserve a good working relationship with Yugoslav Government, legal and aviation officials and above all, impress them with the single-mindedness of their purpose. They had done this with credit: and meanwhile had, as noted above, satisfied themselves as to the efficient performance of both aircraft. Given that both pilots had conducted their flights

[13]*Daily Telegraph* 15 September 1976.
[14]The *Observer* 19 September 1976, *Sunday Mirror* 19 September 1976, *Financial Times* 20 September 1976.
[15]The *Sunday Times* 19 September 1976.

correctly – and the investigators had already gathered some indication of this from the raw tapes of the Zagreb control transmissions which they had been permitted to hear on the 12th – then everything did indeed 'point to a traffic control error'.

None of these suppositions could be conclusive, for the decoded and transcribed evidence of the flight recorders and cockpit voice recorders from both aircraft was yet to come. But there was sufficient now, to justify a joint statement by the British and Yugoslav Governments on the progress of the crash investigation so far. It was issued during the weekend of 18/19th and was reported in terms which could only complete the wretched isolation of the controllers.

> The pilots involved in the ... collision ... have now been all but officially cleared of any blame.
>
> After considerable diplomatic pressure from both Britain and West Germany, the Yugoslavs have agreed to the issue of a joint statement which confirms that the two ... jets ... were operating under instructions from Zagreb air traffic controllers and had been in constant radio contact with them.
>
> The statement also confirms that the responsibility for the collision must rest with at least two of the five ... controllers who were on duty at the time.[16]

To endorse this the newspapers offered 'two key points' from the joint statement:

> The Trident was in level flight at 33 000 ft. and from the time it entered Yugoslav airspace it maintained constant radio contact with the area centre at Zagreb.
>
> The DC9 was cleared to climb after take-off in the direction of the Zagreb VOR beacon ... to a flight level of 35 000 ft. and also maintained constant radio contact with the area centre at Zagreb.[17]

Only *The Times* provided a cautious leavening in describing the alleged error in air traffic control procedures as 'almost certainly the cause of the ... collision'.[18]

It could not, however, forbear from repeating that 'because the accident investigators had "found nothing wrong with either airliner before the collision ..." the inference which the aviation industry is now drawing strongly is that the DC9 was, through a serious oversight on the part of the air traffic control centre, sent directly into the path being flown by the Trident.'[19]

[16]*Daily Telegraph* 20 September 1976.
[17]*Daily Telegraph* 20 September 1976.
[18]*The Times* 20 September 1976.
[19]*Ibid.*

The Times report went on to mention, briefly, the possibility that the error may have been due to a 'malfunctioning radar'; but in the absence of any amplification of that phrase only the 'inference drawn by the aviation industry' remained.

It was left to another newspaper that day to offer an equally strong, but hitherto unpublicised 'inference' by 'aviation experts [who] fear that a "revenge" trial could do more harm than good... if air traffic controllers do not report misses through fear of prosecution, loopholes in safety procedures may never come to light. One tragic outcome of the Zagreb crash is that controllers may just keep quiet after an "air miss".'[20]

One British official said 'Zagreb was a terrible tragedy. But putting people in prison will not bring a single victim back to life.'

It was the first hint of the profound repercussions for civil aviation which would arise from this event.

[20]*Daily Mirror* 20 September 1976.

Chapter Fourteen

In September the city of Zagreb would normally be diverted by the excitement of the largest machine and consumer goods fair of the year. Its newspapers would legitimately exult in the presence, once again, of buyers and exhibitors representing 54 countries and the great influx of visitors whom this event would attract.

Instead, this year, some two thousand people came to Zagreb to mourn.

Three hundred relatives and friends of those who had died were brought in by special flights from Germany, Britain and Turkey, arriving to a sudden silence in the airport's halls and lounges as background music was stilled in respect.

One hundred and eleven people came in on the British flight, among them three priests and 20 stewardesses and aircrew who wished to remember their colleagues. This group was accompanied by Sir Frank McFadzean and Roy Watts, making his second sad pilgrimage to Zagreb that week.

Fifty-three came on the special flight from Istanbul; others who arrived in Zagreb with the 133 relatives of the 107 German victims, included the Under-Secretary of State Heinz Ruhnau of the Federal Ministry of Transport, Inge Donnepp, Minister for Federal Affairs in North-Rhine Westphalia, and Roman Herzog, the Under-Secretary of State from Rheinland-Pfalz. The Suffragan Bishop, Dr Josef Plöger from Cologne and the Protestant member of the High Consistory, Ludwig Quaas also accompanied the bereaved.

An official party led by the British Ambassador, Sir Dougal Stewart, and Yugoslav officials stood bareheaded in pouring rain to greet the passengers as they came down the gangways: their arrival was witnessed by Gunther Braun of the *Kölner Stadt-Anzeiger*. 'Most of them appear composed. But faces which look as if they had turned to stone and red-rimmed eyes bear witness to the suffering of the past days and nights. Some of them are helped along as they enter the airport building and then, a little later, board the seven coaches waiting for them.' Guards of honour, of soldiers and militia saluted as they left the airport.

It is fifteen kilometres to the 'Vatroslav Lisinski', the largest concert hall in a new area of the town and it is here that the funeral service will take place. The journey passes in silence: most of the travellers are occupied with their own thoughts, and the bus drivers, too, have turned off the otherwise customary music.[1]

There could be little solace for these mourners but whatever could be done for them by the Yugoslavs had been done with sincerity and dignity.

Inside the hall the flags of the eight nationalities of the dead hung from the ceiling. The platform was draped in black cloth and wreaths from each of the Governments involved and from friends surrounded the platform on which sat the State Philharmonic Orchestra. The conductor, Vladimir Kranjcevic, had been urgently recalled from a tour.

Some 1400 Yugoslavs, Government leaders and diplomats attended the 40-minute ceremony. Dr Anton Vratusa, the Vice President of Yugoslavia led the 2000 mourners as they stood for one minute in silent meditation, and afterwards, put simply what was in every heart. 'It was a fateful day of terrible coincidence for the two aircraft. Because of a tragic fate these holidaymakers, these bringers of friendship between nations, have remained here for ever. We share with their relatives a mutual agony.'

In due time the buses again took up their charges and, as hundreds lined their route, drove in convoy to Zagreb Cathedral. Here, again, was the grave nobility of a service to bind together those of different creeds: here, again, were wreaths, now piled before the altar. Those who were most closely affected bowed their heads and tried to contain their grief in the quiet rites and the music: thereafter, they knew, they would need to salvage composure from their own thoughts and their own strength. It would be much harder.

But it was clear that, as Dr Vratusa had said, the agony was a mutual burden, for at seven forty-five that evening, thousands of Yugoslavs, forewarned by the Press and now drawn by the cathedral bells, gathered in a display of emotion and fellow-feeling which could not be mistaken. Singly, in couples and groups and throngs they made their way to the cathedral: taxi-drivers left their cabs and joined the farmers and families who had come in from the villages to be led by Archbishop Franjo Kuhariz in a common act of witness.

For this requiem there were five thousand people: they came because it was the only way to enter into the sorrow of this time and in some small degree help to bear a little of the pain.

The British Consul took care of his own nationals until it was time for them to return home: and later that day, the villagers of Vrbovec and Gaj plodded across the sticky mud of the maize fields to pay, in their own fashion, a tribute as impressive as that of the city. They had brought with them all they needed: and for a few hours the poor wreckage at each of the crash sites became a shrine as lighted candles burned about the broken machines.

[1]*Kolner Stadt-Anzeiger* 16 September 1976.

Chapter Fifteen

For Roy Watts and for the British and Yugoslav members of the investigation teams, the end of the funeral ceremonies meant the resumption of their own task.

During that first visit to Zagreb and at the crash sites, Watts – like the others – had encountered the full and utterly confusing flood of personal impressions, hearsay and often totally presumptuous 'explanations'. It was time now to put most of this aside. It was not entirely possible to put aside any of the sense of emotion and outrage and it was not yet time, too, to discard anything, even the most unlikely proposition, of those which had so far been offered. For now, they could be recorded and filed away in the mind: it was all information of a kind, to be thinned down in time by the laborious paring away of fact.

Watts had threaded his way through all of this, noting without comment the allegation by a Cologne Airport spokesman 'that he, in turn ... had been told by counter personnel handling the flight that it had been diverted to Zagreb because of an unspecified technical fault.'

As quickly, the story had been denied by Grimex, the West German representatives of Inex-Adria, but it was merely one among a positive barrel of red herrings. It was wiser, too, not to comment on the statement by an anonymous 'official' that 'neither plane had taken off or touched down at Zagreb so it must have been a pilot's error. We suspect that the pilot in one of the planes was off course.'[1]

The logic behind that suspicion was certainly difficult to follow and indeed, the suggestion received short shrift from Joze Medog, Director General of Inex-Adria. *His* pilot, he claimed, 'could not have been at fault because at that altitude he would have been completely under the control of the Zagreb Tower'.[2]

But it was 'almost certain', too, according to another report,[3] that the British Airways Trident was not to blame in any way for the accident, and it was, said the *Sunday Times* of 12 September, 'almost inconceivable that the accident could have happened if the radar equipment had been working properly. But the secondary radar system at Zagreb has not been officially approved. It has been on test for two years – and still is, thus raising doubts about its reliability.'

[1]Reported in the *Sun* 11 September 1976.
[2]*Daily Mail* 11 September 1976: 'Tower' is here confused with 'Centre'.
[3]*Evening News* 11 September 1976.

In the immediate aftermath of the crash it had seemed, for a while, that logic had indeed foundered, for how else to regard the reported 'possibility' that the Trident 'could have risen a few thousand feet because it had used up a lot of fuel'?[4] That one at least paid a remarkable tribute not only to its author's ignorance but, in addition, to the valour of his employers in publicising it so widely. Yet it was perhaps less worthy of contempt than the selective juxtaposition of general 'information', personal statement and bland implication employed by those of the Press wishing to bolster their own theories with the authority of Roy Watts and anyone else of similar standing.

It was typical of this that within hours, the fact of the crash appeared to demand the injection of the element of pilot-heroism: were there not, after all, the statements of Zivko Koretic and other villagers on which to found this and was there not the detailed eyewitness account by Captain Joe Kroese?

> We saw the other plane, which must have been the British Airways Trident, overtake us. It was a perfectly clear sky and after Villach in Austria it was quite easy to see its condensation trail. While approaching Zagreb I saw a flash and an explosion. The trail of condensation broke and I saw two planes fall. At first we thought one of them was a Yugoslav fighter. But when we came closer we saw the two planes more clearly. At the time of the impact we must have been five miles away. What seemed to be the smaller plane was spinning out of sight. The Trident was tumbling more slowly. *It is possible that the pilot was trying to control it.* My co-pilot thought part of the starboard wing was missing. I immediately radioed Zagreb air traffic control; at first they did not appear to understand . . . There seemed to be utter confusion in the control tower.[5]

'*It is possible*', Kroese had said, '*that the pilot was trying to control it.*' It was the single sentence which registered both in the mind of the public and in editorial offices and it was grasped with an unconcealed eagerness to provide colour and what journalists know as mileage: it required only the ornamentation of the merest hint of agreement from a credible source.

David Montgomery's own comment at the crash site was reported thus: 'Last night it became clear that the pilot of the Trident . . . fought desperately to save the plane.' The British Consul in Zagreb, David Montgomery, said at the scene of the crash: 'The plane, although crushed, was substantially intact and landed on its belly. This suggested to me that the pilot had tried to perform a miracle.'[6]

It was of help in sustaining the drama and the public myth but David Montgomery, of course, had no standing whatsoever in civil aviation. The suggestion of heroism was given much better reinforcement in the *Sunday Express* of 12 September:

[4]*Ibid.*
[5]The *Observer* 12 September 1976.
[6]*Daily Express* 11 September 1976.

After inspecting the wreckage of the Trident, the British investigators believe that the pilot ... fought desperately to land his plane after it had been crippled six miles up. The Trident came down on an even keel with little or no forward movement, and pancaked into a maize field ... Both wings were apparently intact at the moment of impact and the nose section, containing the cockpit, broke off and rolled 300 yards.

This is consistent with Captain Tann trying to stall the plane just before impact and pulling the nose up so that the tailplane struck the ground first ... As the pilot stalled the Trident and the tail hit the ground, the rest of the fuselage would then have slapped down hard into the field; explaining how it burst open underneath and the cockpit sheared off.

Lest this *mélange* of fact and assumption failed to convince, the report continued with further premises: these were so loaded with gruesome import as to decisively quell any thought of demur. 'The British team are trying to find out if the hands of the dead Captain and his First Officer ... were broken. This would indicate that they were firmly gripping the control columns and wrestling with the aircraft at the time of the crash.'

Zivko Koretic, the villagers of Gaj and Vrbovec, Captain Kroese and now, the investigators. Roy Watts's own turn to reflect this illumination came in another report, the same day,[7] which first quoted Captain Kroese: 'he got the impression that the British pilot was "fighting for control" as the Trident plunged towards the ground' and then the British Airways Chief Executive: 'Mr Watts believes the Trident's skipper made a split-second decision to put the plane down in a muddy maize field to soften the impact.'

Sunday's news is notoriously prone to subsequent qualification; readers may not, therefore, have been too confused the following day to find Roy Watts widely reported as subscribing to an entirely different belief: 'Mr Watts said that the medical evidence showed that the victims almost certainly died instantaneously and he discounted theories that the pilot wrestled with controls after the collision ... It appeared that they died as the airliner collided at a closing speed of 1,100 m.p.h. and during the explosive decompression that followed'.[8]

On that point, at least, Watts was satisfied. He had seen the wrecked Trident for himself and could have harboured no doubts about the lethal effect of the DC9's wing on the British pilots: but that intuitive knowledge was now supported by the professional observations of medical specialists, among them Dr Anthony Turner of British Airways and two senior pathologists from the Royal Air Force, Group Captain Balfour and Squadron Leader Tony Cullen.

It was unnecessary therefore, and thankfully so, to pay attention to such further dramatic scenarios as the 'Jet crew's grab for life' also reported that Monday morning.

[7]*News of the World* 12 September 1976.
[8]*Evening News* 13 September 1976.

Captain Dennis Tann and his two first officers had been trained to react immediately to decompression by snatching down oxygen masks above their heads in the cockpit. Trident operations specialist Captain Peter Harper said: 'I believe this is what they would have done immediately as an instinctive reaction.'[9]

Perhaps. But that disintegrated flight deck and the frightful mutilation suffered by its occupants told a very different story and after Watts's firm dismissal of the idea that day, the mileage on heroism ran out.[10]

It did not greatly matter. It would, in any event, soon become yesterday's copy and so be forgotten, along with the human interest stories and the messages of genuine sympathy from the Queen and the Prime Minister and the Chairman of British Airways: it would go into the newspaper archives along with the theories and the maps and the reports of the £3-million 'fastest-ever' insurance pay-out by Lloyd's of London 'within minutes of the crash'.[11]

The matter of insurance would not so easily be put to rest. It represented a notoriously difficult area of contention and, while British Airways were to offer 'up to £25,000 compensation without argument'[12] to relatives of the dead passengers – provided that they can show they have effectively lost that amount (... there will be no automatic liability, our airline spokesman emphasised yesterday) – British and international underwriters were bracing themselves for the claims and litigation which would surely follow. The *Financial Times* of 14 September reported:

> The insurance bill for the 176 victims of the collision ... could be substantially greater than the £11 million suggested.
>
> While the Trident passengers are automatically insured for $45,000 (about £28,280), it has yet to be determined whether any of them had taken out additional insurance or whether the organisations they represented will seek damages for their death. Past experience ... is making the London insurance market exceptionally cautious.
>
> The long litigation over the Turkish Airlines DC10 accident near Paris ... has taught the market that passenger liability suits can be long drawn-out and expensive, especially where US citizens are involved.

---

[9]*Daily Express* 13 September 1976.

[10]Autopsies on crew members were carried out at the respective national forensic establishments: in Britain, the Department of Aviation and Forensic Pathology, Institute of Pathology and Tropical Medicine, Royal Air Force, Halton, and in Yugoslavia, the Institute of Forensic Medicine and Criminology, Medical University of Zagreb.

The three men of the British flight deck crew were found to have been killed in the collision. Their bodies were not complete.

It was only possible to identify Captain Krumpak from the remains of his uniform. Because of its condition it was not possible to analyse the body of the co-pilot Ivanus.

[11]*Financial Times* 13 September 1976.

[12]*Guardian* 14 September 1976.

It was all part of the reality. Other facts lay somewhere to be worried out and pondered and fitted into place by the investigators; but now they had made a considerable advance: they had recovered the flight recorders and the cockpit voice recorders from both aircraft and could attempt to follow the critical moments of their flight.

Chapter Sixteen

Of all the uncertainties facing the crash investigators, one area at least could be approached with confidence: guidance lay within those devices – the flight recorder and the cockpit voice recorder – which the public knew as 'black boxes', albeit that they were in fact red for ease of identification and egg shaped the better to provide protection against impact.

The discovery of these units among the wreckage of both aircraft would enable the investigators to reconstruct the performance of the machines and, in addition, to hear the conversation of the pilots between themselves and as they spoke to the air traffic controllers.

The Daval recycling wire recorder installed in the tail section of the Trident was capable of recording flight parameters, aircraft manoeuvres and engine parameters for a total of 25 hours. It had recorded every second of the flight from the start-up at Heathrow, the radio and standard barometric pressure, altitude, airspeed, axial movement of the aircraft, movement of flaps and rudder, magnetic heading and the engagement of the autopilot. The computer print-out of this information would therefore be of prime importance in establishing the correct functioning of the aircraft.

Roy Watts had seen the Trident's 'black box' for himself: 'when I went to the site, the flight recorder was clearly visible and in moderately good condition. The cockpit voice recorder was found underneath the wreckage and was also in good condition. However, the quick-access cassette, which duplicates the latter part of the black box recording has not been found.' Watts explained the manner in which the investigators would use the recorders: 'Sound tapes were made by the Zagreb air traffic controllers. There are separate tapes for the DC9 and the Trident. Over the last 20 to 40 minutes both aircraft appear on each tape. There is also a third tape made by the Lufthansa captain . . . when fitted to a precise time scale, these tapes will be valuable evidence.'[1] As indeed they were to prove: but first it was necessary to obtain the Trident's flight recorder and cockpit voice recorder from the Yugoslav authorities for immediately after their recovery, on 11 September, the units had been sent to Belgrade.

The natural anxiety of the British investigators for an early opportunity to evaluate the recorded evidence was nevertheless constrained by their position

[1] *The Times* 13 September 1976.

as visitors to the country, and although it was reported on the 20th that 'an invitation to have the various flight recorders decoded in Britain has now been accepted by the Yugoslavs',[2] the interim had seen some restiveness on the part of the Press. Thus, it had already been noted that the British investigators had been permitted to hear the raw tapes from the Zagreb tower (sic) only 'after 48 hours of persuasion'.[3] It had been questioned, too, whether 'the Yugoslavs have the technical facilities for interpreting the highly complex flight recorders'.[4]

It was, therefore, in a somewhat similar vein and under the heading 'Yugoslavs "seal" vital evidence on air crash' that on 14 September the British public were informed of the delay. The *Daily Telegraph* stated:

> The British Airways investigators and those from the Department of Trade now need the recorders to confirm details of what they have seen at the two wreckage sites and what . . . they have heard on the Zagreb recording of the conversations between the pilots and the Yugoslav air traffic controllers.
>
> The investigation teams are not expected to have an opportunity of transcribing what is on the recorders until the end of this week or sometime next week.
>
> Officially, members of the airline and the Government teams in Zagreb say they are happy with the cooperation they are getting from the Yugoslavs and pleased with their 'detailed' investigation.
>
> But, unofficially, one of the British team said yesterday: 'We have reached a stage where we need more information. What we have learned since we arrived in no way shakes our confidence in the (Trident's) crew or the aeroplane.'

Whatever dark implication this 'unofficial' quote purported could not, however, survive for long in the face of the patent integrity and wholeheartedness of Yugoslav co-operation. Within days the recorders had been handed over to British Airways: Yugoslav investigators would come to the United Kingdom to examine the results of the decoding process and both teams would hear the transcripts of tape recordings of the voices of the pilots and the controllers.

The Trident's flight recorder revealed no significant departures from the desirable parameters for that flight.

Cleared for flight level 330, the aircraft was in fact flying at a *recorded altitude* of 33 380 feet. Given a total altimeter error of −420 feet, the *corrected altitude* – the actual height at which the aircraft had been operating – was 32 960 feet. A

<hr />

[2]*Daily Telegraph* 20 September 1976.
[3]*Daily Mail* 13 September 1976. Press reports referred to the air traffic control 'tower' instead of the Area Control Centre.
[4]*Evening News* 13 September 1976.

similarly acceptable tolerance applied to the aircraft heading: the heading for the Upper Blue Five airway was 120°: after crossing the Yugoslav-Austrian border on course for Zagreb VOR and until two minutes and 50 seconds before the collision, the Trident had maintained a heading of 120–122°. Two minutes and 50 seconds before the collision the heading became 115° and for the last five seconds, 116° until the moment of impact.

The aircraft had maintained an IAS, or indicated airspeed, of 291–295 knots, corresponding to a TAS, or true airspeed, of 479 knots, or 905 kilometres per hour. Taking into account a cross-wind component at an angle of 90–100° the aircraft's ground speed was 489 knots.

The autopilot was engaged and it appeared from the characteristics of other parameters that the height lock, which would maintain the aircraft at the selected flight level, was also engaged.[5]

In its turn, the cockpit voice recorder confirmed the essential serenity of that flight. Brian Helm's chatter concerning the discomforts of the route, the crossword, the accident to the helicopter and the bargains at the vegetable market were duly registered, as were Tann's terse but by no means unfriendly replies. There had been no conversation between the crew members and no radio/telephone communications for the last five minutes and 20 seconds before the collision: there had, therefore, been no distractions on the flight deck.

The moment of collision was not recorded: it was concluded that as 'the cockpit was cut through ... the crew suffered fatal injuries'.

The Sundstrand flight recorder from the DC9 was decoded at the Yugoslav Air Transport Technical Centre, and this data, too, confirmed the ordered competence of the aircraft's flight. Like the Trident's own unit, this recorder operated on a second-by-second timebase and also began its work on start-up – in this case, at Split Airport.

The DC9 had been flying at a recorded altitude of 32 445 feet: after adjustment for calibration correction and read-out tolerance this produced a corrected altitude of 32 900–33 050 feet. From Split VOR the aircraft had maintained a heading of 359° to +005°. The heading changed to 353° as the aircraft overflew the non-directional beacon at Kostanjica and this remained constant until the collision.

At that moment the aircraft was in level flight at an indicated air speed of 261 knots. True airspeed was 430 knots and ground speed, owing to tail and side wind, was 465 knots or 861 kilometres per hour.

Again the parameters were much as had been expected, given the correct functioning of the aircraft and in all significant respects this had now been established. There remained only the possibility that something more could be learned from the DC9's cockpit voice recorder.

[5]All data extracted from the report of the Yugoslav Federal Aviation Administration Aircraft Accident Investigation Commission (Komisija za ispitivanje Avionskih Nesreća Savezne Uprave za Civilno Zrakoplovstvo).

It became immediately apparent that this unit had not operated continuously throughout the flight: some defect in the mechanism had caused a number of interruptions amounting in all to a hiatus of one hour and 25 minutes in the tape record.

The tape had stopped some 20 minutes before the collision: there was, therefore, no record at all of the pilot's conversation during this critical phase of the flight and only the most prosaic detail concerning the earlier phases. Thus, the tape revealed a position report from an unidentified aircraft over an unidentified fix at 08.46 and an estimated time over the next fix at 09.08;[6] a conversation between the crews of JP550 and JP548 on the apron of Split Airport at about 9.40 (35 minutes before the accident) and, at approximately 25 minutes before the accident, a message from Split aerodrome control instructing JP550 to report passing Split VOR beacon at flight level 120.

In operational terms, prosaic indeed: but there was something else.

The collision had jarred the recorder back into action and for 25 seconds after the impact, the voice of Dusan Ivanus, the DC9's First Officer could be heard amid a terrible background of sound.

It is believed, states the accident report, that at the eighth second after the crash the crew realised the extent of the disaster which had struck them and for 12 seconds up to the 25th second, there were noises from the cockpit.

The statement is the decent expedient of compassionate men for what follows might well have been deemed to have no place in that document. This study, however, is concerned with human behaviour: it bears witness, therefore, to the ending of one life among so many others.

Ivanus had lived for a brief minute or so after the crash: long enough to try and take in the bloody destruction about him: long enough to comprehend something of what had happened: long enough to be aware of every second of that mortal and fiery plunge from thirty-three thousand feet: and long enough to sear his farewell on the minds of those who would now hear it.

Impact

'. . . JOJ, JOJ, JOJ, JOJ, JOJ!' (a cry of despair)
Unintelligible . . . 'plane', 'fire' or 'help' –
'Gotovi smo, JOJ, JOJ!' (we are finished!)
'Adio, adio.' (Goodbye, goodbye)
'Adio . . .'
'JOJ, JOJ. . . .'
'VATRA!' (fire!)

At the twelfth second there is the sound of the aircraft breaking up. The noises continue up to the twenty-fifth second.

The tape is silent.

[6]The report uses Greenwich Mean Time throughout. Zagreb local time is +1 hour.

Chapter Seventeen

At 8.00 a.m. on 11 April, seven months after the crash, the trial of the accused controllers opened before Judge Branko Zmajevic and the five members of the Grand Council of the Zagreb District Court.

Tasic alone had remained in custody for the whole of that interval: there was, therefore, the aura of an oppressed and burdened spirit to heighten the contrast between the defendants so lucidly drawn by a British newspaper correspondent.[1]

Some brief reference to that description has already been made: it can be seen now in the proper context of a trial to determine the collective survival of this group and as a reflection, so early in the day, of the furious and pitiless alignment which would be brought to bear on one of its members.

Of Gradimir Tasic, then who 'lived with his wife and child in a tiny room in an abandoned airport building'.

> [This] symbolises the gulf separating him from his colleagues. The other flight controllers, who by Yugoslav standards, are remarkably well paid, all live in smart flats in the city. By their own accounts, they enjoy each other's company, go to the same bars and night clubs, and generally lead a gregarious life . . .
>
> Tasic, on the other hand, comes across as a more complex personality. Introverted, highly strung – all these adjectives have been used to describe him by lawyers and journalists attending the trial.
>
> Essentially he is a loner, a man whose main interest is his work. Even his clothes set him apart from the others: he wears an open-neck shirt and a crumpled blue uniform while his colleagues appear in court in well-cut suits . . .[2]

That gulf, so plainly to be observed, was also to manifest itself during a desperate struggle for self-preservation for, while each of the accused would strive against the weight of evidence for a foothold of credibility, those attempts could only undermine their comrade, Tasic. It was less generally obvious, although by no means unremarked, that these struggles, and their outcome, would also inflict a lasting psychological scar on the whole of the air traffic control profession.

[1]Michael Dobbs reporting in the *Sunday Times* 17 April 1977.
[2]*Ibid.*

Albeit diminished, since it had been largely overlaid by the more dramatic aspects of the disaster, some warning of this eventuality had been presented to the public in such phrases as 'revenge trial' and 'it won't bring anyone back to life'; but the real issues were clearly evident to members of the professional bodies within civil aviation.

In some terrible way the technology of the system had failed them: and if, as appeared likely, there had been no wilful or malevolent act to so fatally confound it, then of what use to destroy its servants? And if controllers could be destroyed in an act of mere sacrifice or propitiation, what considerations would influence them once the lesson of their own vulnerability had sunk home? And what manner of service would follow from those so preoccupied?

There were many concerned with this trial who saw the issues in a different and more simplistic light and who were motivated by more tangible frames of reference: thankfully for aviation, however, there were others for whom this trial would represent quite another kind of conflict, namely, that which was epitomised by those questions.

It was also coincident with this trial that, although the air transport industry had hitherto been slow to articulate (and on occasion tragically unable even to recognise) these ideas, the whole climate of enquiry and the open discussion of problems was inexorably coming into being. From the pioneering work of Paul Fitts[3] to the contemporary studies of Stanley Roscoe,[4] those who laboured to analyse human behaviour and its relationship to aviation technology had seen little in the way of popular presentation: nor had aviation benefited from the shibboleths of a misguided sense of professional discretion which – as in medicine – had for so long discouraged the wider availability of 'professional' information and, indeed, even pressed for the suppression of professional evidence offered during legal proceedings.

All this was increasingly under siege: for the previous 10 years had seen the immense strengthening of the 'consumer interest' and the recognition of the right of the public to demand – and get – an accounting. In aviation, therefore, there had already been a small number of remarkable and, occasionally, wholly courageous books (since certain authors risked the displeasure of their employers) which laid bare the fragility of the aforementioned presumptions. Pilots themselves had thrown a searching and unfavourable light on aviation safety: Arne Leibing in *Securite Aerienne*[5] and Captains Vernon W. Lowell in *Airline Safety Is a Myth*[6] and Brian Power-Waters in *Safety Last*.[7] The theme of the need to take practical, instead of merely punitive, steps in the face of a deficient technology was also pursued in Stanley Williamson's *The Munich Air Disaster*[8]

[3]in Roscoe, Stanley N. (1980) *Aviation Psychology*. U.S.A.: Iowa State University Press.
[4]*Ibid.*
[5]Leibing, Arne (1968) *Securite Aerienne*. Paris: Robert Lafont.
[6]Lowell, Captain Vernon W. (1967) *Airline Safety Is a Myth*. U.S.A.: Batholomew House.
[7]Power-Waters, Captain Brian (1974) *Safety Last*. London: Millington.
[8]Williamson, Stanley (1972) *The Munich Air Disaster*. London: Cassirer.

and in Ronald Hurst's *Pilot Error*.[9]

Slowly, such books were beginning – slowly, slowly, but *beginning* – to make some impression on an aviation establishment – operators, administrators, pilots' associations – for the most part unfamiliar with such challenges. They had also contributed to the critical thinking of knowledgeable and experienced private pilots, one of whom, Richard Weston, attorney for the family of the British Airways stewardess, Ruth Pedersen, was now present at the trial. His official capacity, under the Yugoslav legal system, was that of co-prosecutor alongside the State's attorney.

Behind an amiable briskness Weston masked an intense private grief: Ruth Pedersen had been a friend whom he had driven to Heathrow in order to join that last flight. They had made a light-hearted farewell as she entered the terminal building and that had been the brutally casual end of their relationship. Weston had next seen the name of Stewardess Grade 1 Ruth Weinreich Pedersen in the casualty list.

Now he would do what he could for her parents: see to whatever provision was theirs by law – no recompense, he knew, for the bright light of their daughter: but it was all there was for them: or for any of the others.

He sat looking about the unpretentious pale-blue painted courtroom – 'like any suburban Town Hall', he said afterward, save for the huge portrait of President Tito which hung behind the Judge.

On Weston's left was the Deputy Public Prosecutor, Slobodan Tatarac. Beside Weston were other lawyers – those for the airlines and for the families of victims – while to the right was the small enclave of accused men, seven of them grouped together and entirely filling the row of seats: Pelin, thus separated, sat alone across the gangway.

Only Tasic, in profile, furthest away from Weston, sat with a police officer at his back. On his right before two tables were the defence lawyers and a single typist at her own desk. Nearby, and before the screened end wall was the bench from which the Judge, an associate Judge and three jurors commanded the room. It stilled, quite suddenly, as Tatarac, in response to the Judge's gesture of invitation, began to read the indictment aloud.

[9]Hurst, Ronald ed. (1976–8) *Pilot Error*. 1st edn. London: Granada.
Hurst, Ronald and Hurst, Leslie eds. (1982) *Pilot Error*. 2nd edn. London: Granada.

Chapter Eighteen

... the accused are hereby indicted under the Penal Code of Yugoslavia, Articles 271–73, as persons ...
... who by endangering railway, sea or air traffic threaten the lives of men or ... property ... and therefore may be sentenced to a term of imprisonment with hard labour for up to five years ...
... who cause the above by negligence ... one year ... who may be responsible for the supervision of such traffic and who by ineptitude cause the above ... shall be sentenced to imprisonment with hard labour for five years ...
... who cause injury ... ten years ... death ... fifteen years ...
... or great loss of life ... twenty years. ...

In the case of the first accused, Tasic Gradimir – son of Doachin and Stoyana Borgmiladavitch. Born on 29 April 1949 in Nece, Serbia. Living at Vale Camlacha, Vale Cargorcia. Serbian nationality and citizen of Yugoslavia. A qualified Flight Controller, having completed his schooling and training. Married and the father of one child. He has never been sentenced for anything in the past ...

He was imprisoned on 10 September 1976 at 14.00 hours. On 10 September 1976 Tasic Gradimir was operating as a Flight Controller in the upper sector ... The indictment against Tasic reads as follows:

He was acting contrary to the rules, Part 25 of Air Traffic Control, which demand a minimum of separation as a priority; and contrary to the rules of Chapter 4 of the same rules, he accepted co-ordination from the middle sector, although at the same time he was overloaded ... and he did not ensure that the minimum separation was available between the planes DC9 JP550 and Trident BE476.

Patiently, Tatarac detailed the episode:

Tasic had established radio contact with the Trident just after its call at 11.04 12″ and at 11.04 19″ had been told that the Trident was at flight level 330, estimating Zagreb beacon at One Four.
... 11.14 04″ Tasic established radio contact with the DC9. At 11.14 10″ he was informed that the aircraft was passing flight level 325 and estimated Zagreb VOR at the same time, i.e. One Four, as the Trident ...

It is charged that he was aware of the danger: (but) instead of taking urgent avoiding action proceeded with normal communications with the DC9 pilot. At 11.14 14″ he requested the DC9's flight level in the Croatian language and at 11.14 17″ received the message that 'the plane was climbing and was passing 327'. After a further five seconds, at 11.14 22″ he asked the DC9 to maintain his present level and then asked the DC9 to contact him when he passed the VOR. The DC9 pilot then asked Tasic which flight level he was talking about ...

At 11.14 29″ Tasic replied that the pilot should hold the level through which he was climbing, because he had a plane in front of him at level 335 from left to right: although he knew that this plane was at flight level 330. At 11.14 38″, the DC9 pilot replied that he was flying at 330. After a further few seconds he managed to get the aircraft in a horizontal plane at 330. (But) because Tasic had not ensured the minimum separation between these two planes, and had not foreseen in sufficient time, the crash situation that was about to occur; and because he was using imprecise air traffic measures to avoid a crash situation, the two collided ... at 11.15, overhead the Zagreb VOR.

Wooden-faced, the controllers stared ahead as Tatarac described the moment of impact, the disintegration of the two aircraft and the location of the wreckage ... near the villages of Gaj and Dvoriste ...

One hundred and seventy-six lives.
DC9 insurance, six and one half million dollars ...
for the passengers ... two million, one hundred and sixty thousand ...
for baggage ... eighty-six thousand four hundred ...

And for the Trident:

Seven million, two hundred thousand dollars ...
for the passengers: one million eighty thousand ...
for baggage: forty-three thousand two hundred ...

'There is a total loss', Tatarac said, 'of seventeen million, sixty-nine thousand, four hundred dollars.' He paused: ... because of his behaviour Tasic is accused of endangering air traffic: he is also accused of endangering lives and property on a far larger scale which resulted in the deaths of many people, and very heavy losses. He is accused of committing the crime against the common security of persons and property by endangering public traffic described in Article 273, Section 3, and in connection with Articles 271, Section 2, and 273, Section 2, of the Yugoslav penal code. This is punishable under Article 273, Section 3 of the same law.

A momentary pause: then Tatarac went on:
... the second and third accused, Delic, Antere, Head of the Flight Control Service and Munjas, Milan, Head of District Flight Control ... they are accused of not organising the work of the district control according to the rules, especially with regard to the amount of traffic one controller can safely manage, bearing in mind the required separation, and rules and regulations: so that

Tasic Gradimir was overloaded between 11.04 12″ and 11.16 and engaged in continuous radio conversations with 11 aircraft in the upper sector ...

From 11.06 59″ to 11.07 40″ he took over the co-ordination of another aircraft[1] although he only had 40 seconds to consider what it was necessary to do in order to ensure that the aircraft reached its correct altitude safely. It is also charged that Delic and Munjas had not paid enough attention to the discipline of the controllers, which was necessary to ensure the safety of passing aircraft. They were aware from the results of disciplinary proceedings against controllers that they had been careless. They particularly knew that the first accused, Tasic, and the sixth accused, Tepes, had been disciplined ... Tasic had been warned, and Tepes had been warned and fined, because they had been late to work on a number of occasions. On several occasions they had also left their positions without permission from the chief of shift ...

They also knew from pilots' reports that not enough clearance had been given by various controllers, who had been taking short cuts and had been simplifying procedural rules. They also knew that the fifth accused, Hochberger, had been disciplined with a warning, and the sixth accused, Tepes, had ...

Frowning, Weston began to make notes. It was no surprise to him that the defendants should be assailed in such comprehensive fashion since it was clear that the State's case was in competent hands: and in that sense, his own position as a special representative could only be strengthened by any finding against these men. There seemed every likelihood that that would happen, too: some, if not all of these men would be made to pay for what had happened –

It wouldn't bring anyone back to life.

Weston thrust the thought aside and brought himself back to Tatarac's unsparing catalogue.

... been punished with a public warning, because he had been cutting short procedural clearances. In fact, separation was too fine or was not wide enough, or Tepes had aircraft flying too low and had used radar when it was forbidden, or he had been allowing aircraft to descend too close together on the final approach. He was fined because he allowed the aircraft to leave the airway ... when he ignored the rules.

Item by item Tatarac diminished the two chiefs, piling the short-comings of their men on to their own ... they did not take adequate measures to ensure that the ICAO rules and regulations had been obeyed, although they are additional to the Yugoslav rules. They also knew from the pilot's report and other sources that cases of endangering aircraft safety and security had occurred quite often, and they also knew from the results of disciplinary hearings in Zagreb District Control that Tasic, Hochberger, Tepes and Pelin had been called on more than one occasion before a disciplinary committee because they had caused near misses between aircraft. Tasic had also endangered air traffic to such an extent that there had been the possibility of a crash between aircraft Yankee Uniform 374 and Transavia HV272 ...

[1] JP550.

90

Behind the controllers, the group of international journalists worked away as the first peg on which their stories would hang was driven firmly into place.

... There was a further incident between YU 230 DC9 and YU VHL BE60. The fifth accused, Hochberger, caused near misses between aircraft YU 711 and YU 891, with a further incident between YU 894 and YU 323. Tepes endangered air traffic security by sending the plane YU 222 through a corridor where the planes YU 210 and LZ 1515 were already flying. Pelin caused a near miss between YU 230 and CK 889 ...

Delic and Munjas bore this responsibility, summed the Prosecutor since they ...

... had not found the reasons for these incidents or had not trained the controllers sufficiently to avoid such incidents. They had not examined their controllers in detail before renewing their licences. They had not determined precisely the duties of a procedural controller or of a radar controller. And they had done nothing to avoid a situation where an unqualified controller might end up having to operate the radar control system. Neither did they ensure that the flight controller could instruct a pilot adequately in an emergency situation, or enforced the ICAO rule about the use of the English language for standard phraseology and terminology. The rules demand that tapes should be checked monthly, and disciplinary action should be taken for deviations from official phraseology, but they knew from the transcript of the tapes in the disciplinary case that Tasic had been using Croatian when communicating with aircraft. Also, in the middle of June they did not follow up the proceedings of the disciplinary case previously mentioned ...

Delic and Munjas sat tight-lipped as Tatarac finished. The Prosecutor looked about the Court then returned to the indictment.

The fourth accused, he said: Dajcic Julije, chief of shift ... He is accused of not ensuring that the work of the controllers was performed in accordance with the existing rules of the Federal Authority. He did not ensure that the controllers or assistant controllers remained at their stations before they were officially allowed to go. And, although he could see by looking at the slips on the console, that Tasic was overloaded ... he did not give him assistance.

The fifth accused, Hochberger Mladen, assistant to the upper sector controller ...

His shift ended at 11.00 hours and at 11.05 he left his place of work: he did so without giving any consideration to the possible overload of procedural controller Tasic ... even though it may have been evident from the number of flight slips in the procedural console ... and by the fact that, at 11.04 12″, Tasic commenced continuous radio communication with aircraft.

He left the room to look for his replacement, the sixth accused, Nenad Tepes, and when he found him, made the handover (of his traffic) without familiarising him with the situation as required ...

As for Tepes, he is accused of arriving late for duty so that Tasic had to make four telephone calls to Belgrade between 11.05 11″ and 11.09 55″; of accepting the handover outside the control room; and of failing to familiarise himself with the situation, then and later, at approximately 11.10, when he arrived at his place of work, although he could see that Tasic was overloaded . . .

Fair enough, Weston thought. The onus was being shared so far: it was going to be rough going for all of them.

. . . Seventh accused, Erjavec Bojan, air traffic controller, and eighth accused, Pelin Gradimir, assistant air traffic controller . . .

. . . after the DC9 pilot had reached level 260 at 11.05 57″ he requested a climb to a higher level. They are accused that, having stated that all levels in the middle sector were occupied, they failed to act according to rule 28 of the FAA[2] control rules, which demands that the controller transfers responsibility for control only after he has removed any possible risks of conflict with other aircraft which are being controlled by the person to whom they wish to pass control. Instead . . . seventh accused, Erjavec, sent the eighth accused, Pelin, to Tasic, to co-ordinate the transfer, although they had both seen that the first accused was working alone and was overloaded. Tasic waved his hand to tell them he did not want to be disturbed and in the interval between 11.06 59″ and 11.07 40″, when Tasic had finished his radio communications, Erjavec answered the DC9 pilot. Neither had given the upper sector a slip with the necessary information (according to rules 211A, 203, 212 and 215 of Federal regulations) which would have given a timely warning to the controller. They passed the slip from the middle to the upper sectors only after the DC9 had crossed all middle sector levels – at 11.12 12″, when the plane left the middle sector . . .

A firm headshaking, barely acknowledged by Tatarac's glance, came from the two men. Unhurriedly, the even voice went on:

. . . during this co-ordination Erjavec was using a radar, for which he is not qualified, and at 11.07 he asked Pelin to identify the DC9 on Tasic's radar in the upper sector, because he, himself, was not as experienced as Pelin. The DC9 was at level 260 at that point, and Tasic selected his strata on the Julia combined computer radar screen. But the identification of the DC9 was not complete because the aircraft below flight level 330 had no coding or call sign beside their blips. At 11.07, according to rule 565 of Federal Authority control rules, Pelin made an incomplete radar identification of a target on the radar screen. Thus, the correct radar handover had not been completed in accordance with rules 567, 568 and 569 of the Federal Authority control rules, and the use of radar rules contained in Chapter 10 of I.C.A.O. 4444 document 4, which demands that the minimum separation, position, and change of sector information should be checked and radio contact should be maintained until the transfer is completed. But instead, Pelin thought that the radar transfer was complete, although, according to rules 566C, 531 and 532 of Federal Rules, he should have said 'Stop Squawk, Mode Code' and passed the pilot going to the upper

[2]Yugoslav Federal Aviation Administration.

sector A2300 and 2377 (the appropriate code range), which would have been correct ...

Painstakingly detailed and crushing in its massive demolition, the indictment presented a formidable challenge: plainly, Judge Jakovac's Commission of Enquiry had done its work well. Inevitably, however, this survey of the controllers' past as well as current misdeeds provided the Press with that keynote of official laxity and individual dereliction which overlaid their subsequent reports. To this would be added the further humiliations of material deficiences: and the spectacle of erstwhile colleagues rending each other in their efforts to survive.

It was Tasic, now called to witness, who would inflict the first wounds. He took his place before the Judge and began to speak.

Chapter Nineteen

'I will speak the truth', he said. 'I do not wish to defend myself, but I will speak the truth as the truth speaks for me.' And the truth, Tasic told the Court, was that the fatal errors were committed by the controllers of the middle sector and by the pilot of the DC9, while he himself had been doing the work of three men – controller, assistant controller and radar controller.

Nor, he insisted, had he given the DC9 clearance to climb above 31 000 feet: that had come from the middle sector 'while I was talking to 11 different aircraft and making telephone calls to the Belgrade control which should have been made by my (absent) assistant'. It was a fact, he said, that they were in any case short of qualified controllers at Zagreb flight control: it was no secret that 30 controllers were doing the work of 70.

Tasic now began an account of the critical minutes before the collision. The middle sector controllers, he said, had allowed the DC9 to climb without informing him: they had misunderstood his instructions (that wave of the hand?) . . . and transferred the DC9 to him without giving him the flight details . . . and also, the pilot of the DC9 was about two minutes late in reporting in to the upper sector. And when he said he was already at 32 500 feet and still climbing, said Tasic . . . 'I couldn't believe it. I asked for his altitude and then ordered him to level – in Serbo-Croat – because the Trident was in front of him from left to right'. That was about 40 seconds before the crash, Tasic affirmed. 'There wasn't time to do anything else.'

'I looked at the radar and saw the two blips converge. They moved apart and I thought there had been a near miss: but it was only when I tried to call the captain of the DC9 that I realised what had happened.' Tasic's voice trailed off. Then he said strongly:

There were three things – three misunderstandings – which caused the crash. First, the radar showed the Trident to be at 33 500 feet [but Captain Tann had already informed Tasic at 10.04 12″ that he was flying at 33 000 feet] . . . second, that he had thought the DC9 to be at 31 000 feet until, barely 40 seconds before the collision, the DC9 pilot had declared his aircraft to be at level 327 . . . 'and climbing'; . . . and third, there had been the delayed response when he had instructed the DC9 pilot to maintain his present level and the pilot had failed to understand him. 'What level?' he had asked . . . and by that time the aircraft had reached the same height as the Trident .

Weston watched the suffering Tasic as he groped finally to move the focus from himself and on to another target.

'There has always been poor co-ordination between the sectors at Zagreb', the controller was saying: 'This had been responsible for a number of near misses.' The British investigators had seen this for themselves –.

'Why did you not make more use of the radar?' asked the Judge. 'Because', replied Tasic, 'the set hadn't been properly adjusted since they put in in three years ago. It was unreliable . . . and therefore not used by the controllers.'[1]

Tasic would answer another of the presiding Judge's questions with equal despatch. What he was not to know, or did not immediately realise, was the secret knowledge and the deadly intent which prompted what appeared to be a comparatively innocent question. But after hearing the air traffic control tapes, the Commission of Enquiry had noted that despite Tasic's appreciation of the impending danger he had not revealed any anxiety to the DC9 pilot, or shown any excitement – his voice had remained calm.

'Why was that?'

'It is in the instructions,' said Tasic. 'Controllers are not expected to panic . . .'[2] 'It was right to remain calm': and unspoken was the corollary: it was *possible* to remain calm.

The answer was noted. It would rebound on Tasic at length, with the cruellest of endorsements.

On the second day it was the turn of Delic and Munjas and it was Delic who began his own evidence with the confident assertion that rigid discipline was maintained at the Zagreb Flight Control Centre.[3] He could not be blamed, he said, or even be partly blamed, for contributing to this accident by failing to check the work load of the staff, or failing to discipline the controllers . . .

'In the first place there were no regulations stipulating the maximum work each controller should handle . . .'

'But Tasic claimed to have been overloaded?'

Delic shrugged off the suggestion.

'It is the responsibility of the Yugoslav Civil Aviation Administration to determine the maximum work load for controllers', he said briefly . . . 'My job is just to see that the regulations are carried out.' He turned to the 'difficulties' which they had encountered in the region: yes, there had been 32 narrow escapes from mid-air collisions in the past five years, he said, adding: 'most of the complaints received and acted on were made by Yugoslav pilots . . . it had been necessary to dismiss two controllers . . . for carelessness and lack of training.

'Thirty-two near misses?'

[1]*Daily Mail* 12 April 1977.
[2]*The Times* 12 April 1977.
[3]*Daily Telegraph* 13 April 1977.

'One must remember that in that time (the previous five years), Zagreb had become the second busiest air cross-roads in Europe; and during this period there had been 700 000 flights through the region.'

'Despite that, Tasic says you were short of controllers?'

'... that 30 controllers were doing the work of 70 ...?'

'Well', said Delic in deprecation, 'we – the Zagreb flight centre – needed 40 trained controllers: we had to make do with 30. I should say', he admitted, 'that in this respect we were behind some other European states ... but since the disaster, this state of affairs had been radically improved.'[4]

'What is your opinion of Tasic as a controller?'

Delic answered without hesitation: 'Gradimir Tasic is one of our best controllers: I recall him saving an aircraft from disaster by warning him that the undercarriage was not down as he came in to land at Zagreb. Tasic saw the danger from the control tower ...'

Judge Zmajevic led the questions. Occasionally he would signal the proceedings to halt while he summed up, speaking directly to the typist. Each time, the witness, grimly thankful for the respite, would visibly gather himself for the next bout of questions.

'The technical equipment? Tasic had said that the radar was unreliable: that it was not used by the controllers?'

'It's the same problem,' Delic explained: 'we just haven't been able to keep up with this increase in air traffic in the past five years. Not in the number of trained controllers we need ... not in the matter of technical equipment to guide the aircraft ... we rely very much on procedural control ...'

'Yes, there was a modern flight control system, a Swedish one. We installed this at Zagreb three years ago but we don't use it for separating the aircraft: only for checking on the pilots' position reports.'

'Why is that?'

'Because', said Delic, 'that equipment has not yet been commissioned. In a number of respects it is not yet satisfactory: it fails to meet the requirements we have specified in the contract.'

Delic stood down to be replaced at the lectern by Milan Munjas. Broadly, Munjas supported his superior. No, it was not true that he himself had failed to keep a proper discipline among the air traffic control staff. Yes, it was true that they needed 40 extra controllers at Zagreb. ... a recent study[5] had shown this: it was 'more than double' the number of staff we had there at the time of the crash[6] 'and this is the case today'.

'You say that you did not fail to keep discipline?'

It is in the record, Munjas replied: 'There is almost no case in which violations of the rules were not punished ... in fact, 37 controllers have been

[4]*The Times* 13 April 1977.
[5]Peter Ristic in the *Observer* 17 April 1977.
[6]*Ibid.*

disciplined since 1974 and it was Ante Delic and myself who brought the charges.'

'Disciplined?'

'Disciplined, yes – three of them were sent for further training. We took one man's licence away and demoted him. We demoted others and two were sent back to whatever jobs they'd been doing previously. We took them out of the training school and did that. And we fired one man, too. It's in the record, as I say: so how can it be said that there was no discipline?'

Munjas dealt with the next questions as expeditiously. So far from ignoring the rules, they – Delic and himself – had introduced additional ones covering different contingencies ... What about the rule that controllers should record on the strip the time an aircraft left the flight level? 'A waste of time', said Munjas decisively. It did nothing useful except clutter up the strip with superfluous data.

It is interesting that Munjas displayed a quality which later earned him a tribute rarely extended during the trial: that he was 'perhaps the only defendant in the first week to distinguish himself.'[7]

It is certain that the procession of witnesses in that time provided an adequate yardstick for observation. Munjas had impressed: thereafter, however, swift currents of fear and mutual accusation would submerge completely not only Tasic's 'truth', but also, that of the others. It became by no means sure that it might eventually be brought to the surface.

Nor, on 13 April, the third day of the trial, did Julije Dajcic remain long at the witnesses' lectern. 'I deny the allegations against me', he declared. 'I did not neglect my duty and I did not fail to supervise the work of the controllers.'

'And between 11.00 and 11.15 on that morning, what were you doing?'

'I was doing what I was supposed to be doing at that time', said Dajcic with heavy emphasis: 'I was writing the daily schedule for the shift next day.'

'At your desk?'

'At my desk in the control room, yes. It's part of the job to spend some time there.'

'And you could see the whole room from that position?'

'It is not well lit', Dajcic began slowly. Smiling faintly, he said: 'I'm short. There's a high board at my table and I must stand up to look over the whole room.'

'But you could see the upper sector console?'

'Sometimes: that's on my right – if I turn towards the left of the room, no.'

'Did anyone approach you that morning, at that time, for permission to leave the room?'

'No', said Dajcic. 'No one asked me that.'

[7]Ibid.

'Did Tasic ask for your help on the console: could you see if he was overburdened?'

'Tasic did not ask me that, either. It is not possible to know the position, whether a controller is overburdened or not, just by looking at the strips on his console.'

'Why not?'

'They may not all be active strips in the console. Only the controller knows his current traffic and those aircraft which haven't yet called him . . . if he needs help', Dajcic ended, 'he can ask.'

Chapter Twenty

That there was, and would continue to be contention and hostility between the controllers could not be in doubt; for if there had been no other indication the notes of discord were plainly set down, in those phrases 'it has been impossible to substantiate his statement . . .' and 'there is a conflict of evidence' which occur in the report of the Accident Investigation Commission.

It was Hochberger, followed to the stand by Tepes, who helped to translate these words into scenes of emotional violence: not, however, before Hochberger himself had faced some damaging questions.

> . . . Yes, he had left the console at 11.05 – his replacement Tepes had been due to arrive at 11.00 and had failed to do so. But we weren't very busy, anyway: we chatted quite a lot . . .
>
> Yes, he knew it was forbidden to leave the work position: but Tasic had taken the headphones from him and told him he could go –

'Tasic told you that you could go?'

'Well . . . I went to see if I could help at another position but there was nothing to do . . . so I offered to help Tasic because Tepes still hadn't arrived . . .'

'And what did Tasic say?'

'He didn't answer', Hochberger claimed: 'it was then I went to Dajcic and told him that I was going to find Tepes.'

'But Dajcic denies that you approached him: are you sure he was at his desk?'

Hochberger did not answer directly. 'I told him', he said stubbornly. 'I told him that Tepes hadn't arrived and that I was going to find him . . .'

'But Dajcic hadn't given you permission to leave the sector?'

'During the first hearing, in your evidence to the Commission, you didn't mention asking for permission. Or that you showed Dajcic a watch when you spoke to him? You also said that Dajcic just shrugged his shoulders when you told him? That Dajcic knew there was only one person left on the upper sector?'

Feebly, Hochberger attempted to weather all this. It's possible, he offered, that Dajcic didn't understand me – but other people saw me approach Dajcic's desk.[1] Hochberger felt himself to be on firmer ground when the Prosecutor turned to the question of whether Tasic – as had been claimed in his defence –

---

[1] Hochberger called female members of the control room staff as witnesses. Their evidence was rejected as unreliable or hearsay.

had been 'left alone for a long period and was under pressure'.

'You say Tasic wasn't overburdened?'

He liked working alone, Hochberger affirmed: it's well known, Tasic had insisted, he said, on taking over as flight controller at 10.56; he'd moved into the controller's seat then, even though he was supposed to be working as the assistant controller during that shift ... Tepes will confirm all this ...[2]

'You found Tepes and handed over the traffic. Where did this take place?'

'Outside the control room: that is, at the door.'

'That is against the rules? Shouldn't the hand over take place at the console?' Hochberger rode this out, too. Normally, he agreed, that would have happened. But Tepes had been late – he'd been in the lavatory. 'I found him entering the control room.'

'What time was that?'

'About 11.07', said Hochberger: ... 'I'm sure that Tepes was at work by 11.07 ...'

'Do you know that the Commission report states that the upper sector assistant controller (Tepes) was absent from his post from 11.0510″ until 11.10 . . . and probably until 11.13? During a period when the sector controller's work load was high?'

The questions did not appear to require any answer: bruised, Hochberger was at last permitted to stand down.

'Tepes, Nenad ... you were late in taking over your duties?'

'I did the shift from 8.0 a.m. to 10.0 a.m.', Tepes said: 'then I went to the rest room and read for a while, till about 10.45; then ...'

'Then what?'

'It was nearly time for my next shift. I went outside to get a pencil from my car, then I went back to the control building ...'

'To your post?'

'No', said Tepes, 'not right away. I went to the lavatory. I needed to: I've been having a lot of stomach trouble ... It's my digestion', he volunteered.

'When did Hochberger meet you?'

'I heard him calling me when I was in the lavatory: I met him at the door of the control room at about 11.07 ... well, anyway, I was in my working place by 11.0920″ ...'

'What was the situation when you arrived?'

'Tasic was on the phone. I asked him if there was a lot of work and he said; no, not a lot[3] ... and then I familiarised myself with the traffic in the sector ...'

Tasic's counsel put the next question:

'You've been late on duty before – you've had disciplinary action taken against you?'

[2]*Guardian* 14 April 1977.
[3]*Guardian* 14 April 1977.

'That's so', said Tepes: 'but may I point out that I've also had commendations for my work?'

The final moments of Tepes' evidence provided a novel experience for correspondents of the Western Press, for in a gesture variously described as 'theatrical, a spectacle worthy of a Perry Mason TV trial'[4] and again, as a moment 'when the judge behaved more like a Hollywood director than a lawyer' Judge Zmajevic called on Tasic to stand beside Nenad Tepes.

'Now', the Judge said to Tasic: 'you have heard what this witness has to say. You may question him yourself – are his statements correct?'

'No', said Tasic, 'they are not correct.' With increasing intensity and in obvious distress he challenged Tepes.

'You came to your post at 05 or 06 and went away.'

Angrily, Tepes rejected the charge: 'That's not true. I came to my post at the time I stated in my defence and I didn't leave my post after that.' And, he made clear, Tasic was in the wrong seat – he should have been the assistant controller for that shift.

Tasic was pale, trembling with his own anger.

'We didn't have an agreement that morning about who was going to do which job –.'

'We did.' Tepes pressed the point home: 'I can prove that there was an agreement because between eight and ten o'clock when I worked on the sector, there wasn't a word of mine on the tape. I'd just been monitoring the work and it can be confirmed by the others of the control staff . . .'

The Court watched as the bitter wrangle ended. Beside Judge Zmajevic, and for the third day of the trial in succession, the elderly juror, head on chest, dozed on.

[4] *Sunday Times* 17 April 1977.

Chapter Twenty-one

To Ronald Hurst

From Richard Weston
London,

August 16th

Dear Ronnie,

Since we first began this work together you have asked me to tell you about Ruth and about my involvement in the Zagreb case. I asked you to let the matter rest while we got on with everything else and until I felt better able to write about it. I suppose that that was an evasion: I just didn't know how to start putting it down and I couldn't think of it without a considerable sadness.

However, I've realised that it won't get any easier. I'm, therefore, making the effort now, for the first time. It can all go onto the tape recorder, as I recall it, and then my secretary can type it up. If you feel that there is anything else you want to know we can talk about it.

My involvement in the Zagreb story started in January 1975, when, as a 38-year-old bachelor divorcee, I was walking through the swing doors at the Heathrow Hotel on my way to their health club. I'd got a very bad limp, occasioned by a torn leg muscle. I remember that I held open the door for a striking blonde – I like to think I would have done the same for an old lady! Seeing my stick and discomfort, she smiled sweetly and said 'after you', and I remember replying that 'people in that uniform (she was dressed in a B.E.A. uniform) are always doing things for me: now I have the chance the return the compliment ...'

And that was the end of the meeting. Some months later, I think in March, I chanced to see the same girl walking out of the health club and enquired of the receptionist who she was. They told me her name was Ruth Pedersen, that she was a B.E.A. stewardess and quite regularly visited the beauticians' department. I thereupon wrote a note and said 'next time she comes in, would you give her this?' It was an invitation to dinner.

Apparently, I omitted to put on the note my address or telephone number and she never called. However, she did leave a message for me at the reception desk in May, as a result of which I called her. We struck up a friendship which was to last until her death in Zagreb in September 1976. She was 26 at the time; she had worked for British Airways for four years as a stewardess. She was born in a little fishing village in Denmark, the second of three daughters of the family; that was all I knew of her background. She had been in England for some years and spoke English fluently. She also spoke excellent Danish, Swedish and Norwegian. I do not speak those languages, but certainly, as far as her English was concerned she had no accent whatsoever. She was an accomplished young lady who was actually frightened of flying and when people used to ask her (usually a first-class traveller) 'why do you do this job?' she would reply, 'because if I did not, I would probably be a secretary to someone like you.' She was attractively blunt. She had worked on Royal flights, she appeared on the British Airways television advertisements 'Fly the Flag' and she was sent to California by the airline to be photographed and filmed in the advertisements and brochure for the Tri-Star, which was about to be put into service.

From this record it will be understood that she was meticulous about her appearance and proud of her job. She had only been late or missed a flight three times, I think she told me, in four years and had seldom, if ever, been off for ill health. During 1976 she applied for, and succeeded in getting a job at the Heston Training Centre for British Airways stewardesses – a job that she was due to start on Monday 13th September. During August, on a flight to Tel Aviv, she suffered damage to an ear as a result of which she was 'off flying' until she recovered. The system in the airline in such a case is that you have to report regularly to the doctor who then signs you fit or unfit as the case may be. On Tuesday morning, the 7th September 1976 she had to check in with the doctor and as she left the house she said something which was uncharacteristic of her. 'I hope that he will sign me sick because I have so much to do before I start at Heston on Monday, what with uniforms to clean and press, paperwork etc., and they have given me a book to read "I'm OK. You're OK".' I say this was uncharacteristic of her because she did not usually like to take time off. When she came back later in the day she said, and I quote again because they are words I remember for obvious reasons, 'my luck, he signed me fit; I have got an Athens tomorrow and Paris on Thursday and I finish with an Istanbul on Friday'. It had been my wont during our relationship to telephone the airport to check that her incoming flight was on time, so that I could meet her as I lived quite nearby. On occasion, if she were on a long flight and I was going to be out, I would ask my secretary to telephone (they knew each other quite well) and leave a message on my desk to say whether she was going to be delayed or not.

On Friday 10th, I was due to go to Farnborough and my secretary had asked if she could leave early that evening because she had something she wanted to do, and I said 'yes, of course you can, but would you just check the time of Ruth's return from Istanbul and leave a note on my desk before you go.' I returned to the house at about 6 o'clock and was surprised to see my secretary's car still in the drive. As I drew up she came out and was clearly distressed. I asked what the matter was and she said 'I have got some rather bad news, you had better come inside'. I said 'well, you had better tell me what it is' and she said 'there has been an accident, why don't you come inside?' And I said 'what kind of accident?' and she then told me that Ruth had been killed. I went inside the house and started to dial the airline and my secretary said 'I have already spoken to them', and I said 'I just want to check that she was on board' and my secretary said that she had already checked this. On the evening television news it then became clear what had happened, and by this time, of course, the blame was already being laid on the air traffic controllers, although full details were not available.

The next morning I went to see her supervisor at Heathrow and he was in a state of some distress himself, as most of the cabin crew on the Trident were close friends who had worked on his flight for some time. He told me the airline was planning a 'pilgrimage' for relatives on the Tuesday and I would be welcome to join them if I wished.

By this time I had, in fact, spoken to Ruth's parents (the evening before). Ron, I do not know how much more of this you want? I went down to Zagreb to the service and then went off to Denmark and saw her parents; they asked me to handle everything. I went down to Zagreb again after the identification process was complete, and took her coffin personally back to Denmark and attended the funeral the following day. There is a certain amount of irony behind this, but I do not know whether you want it – let me know and I will give you further details if you think it is necessary. But for the moment I shall move on to early 1977.

By January 1977 I had appointed a lawyer in Zagreb to look after my interests as administrator of her estate, to which I had taken out letters of administration as the attorney of her parents, and also as the attorney of her family themselves. The eight air traffic controllers who had been arrested were due to go on trial in April and three days before the trial the lawyer whom I had appointed died. With the assistance of the British Consul in Zagreb I managed to find another lawyer called Dragomir Modrusan,[1] who spoke fluent English and had agreed to assist.

I had many reasons for being interested in the trial; in the first place I had undertaken a responsibility in relation to Ruth's family and in the second place I was an administrator and therefore bound by my duty to the court in England to find out what was going to happen in relation to compensation. Thirdly, I was a lawyer and the idea of a criminal trial in a foreign country was of interest to me. Fourthly, I am a pilot and particularly interested in aviation law, which is my field of practice, so the idea of air traffic controllers on criminal charges was particularly fascinating, and finally, the case itself was so unusual that it would have been of interest to me even had I been an outsider. The law in Yugoslavia however, provides that where a party has been aggrieved by a criminal act, although the State prosecutes through the State Prosecutor the aggrieved party may appear as a co-prosecutor; and this was the role which I played, along with British Airways and another lawyer, and indeed in an extreme case there could have been 176 lawyers – that is to say that each family could have decided to be represented as a co-prosecutor: in the event, I think only a minority of them did.

[1]Occasionally referred to in text as 'Drasko'.

Finally, the law in Yugoslavia also provides that in a criminal action such as this, the judge has the power to award damages or compensation to the victims or their families. In fact, during the early part of this trial the judge said that he was not going to get involved in the award of civil damages. There were obvious reasons; there were eight defendants, there were 176 victims and it was necessary to deal with the criminal side expeditiously in the interests of justice, and those interests would not have been served by the judge starting to try 176 different claims for compensation. The trial was quite fascinating. The judge who presided was quite young – I believe he was 39 – and he was assisted by five assessors, two of whom I think were pensioners and three who were paid officials. It was a formal yet informal atmosphere; that is to say although everyone wore sports jackets (except the judge, who wore a lounge suit), one had to stand up when the judge came in, one had to be quiet and the questions and answers were dealt with in a dignified way. The courtroom itself was an impressive room, having been the very same room where Tito was tried for communist activities in 1923. It was difficult to follow the trial because Modrusan sat at the Lawyers' bench next to the State Prosecutor and the British Airways lawyer, and I sat immediately next to him, as it were, at right angles. Somebody from the British Consul came with me on many occasions during the five weeks of the trial and although I did not speak any Croatian, with a little bit of help either from Drasko (my lawyer's nickname), or from the British Consul's representative, I managed to follow most of what was going on although obviously I missed some of the finer points and the nuances: but on occasion, when two witnesses disagreed the judge would call them both up at the same time and ask them to argue the point and it was quite clear as to what was happening with accusatory fingers being pointed and voices raised. The defence lawyers all sat on the other side of the room in a row.

The court sat at eight every morning and at a quarter to ten we would adjourn to the local coffee house (nicknamed the Phoney Witness) for thick black coffee, iced water and a bun, before returning to court at a quarter past ten until twelve. The newspapers were already full of the fact that the controllers were responsible for the crash. The issue had been prejudged; only the scapegoat had to be found.

For myself, I arrived in Zagreb fully convinced that the air traffic controllers responsible for this terrible thing should be hung, but sitting there watching the plot unfold I gradually began to have second thoughts. To me, at that time, an air traffic controller was a voice on a radio, usually a very reassuring voice, an authoritative voice, a capable voice, a well-organised voice which I obeyed out of regard for my own safety and out of pride for my own airmanship. Sometimes a controller was a bloody nuisance, sometimes they could seem pig-headed, but usually, one is only with a particular controller for a short time before being handed over. For instance, the ground controller taxies you out and hands you over to the tower. The tower controller takes you off and hands you over to departures. The departure controller hands you over to radar, radar then hands you over to en route. So that one might be with a chap for ten or fifteen minutes, and it is not worth getting annoyed if the fellow is off colour, because you are going to be handed over anyway. But here was something new. These people were actually standing in a criminal dock, one of them being escorted in every morning by the police (Tasic was in custody still, after seven months) – and the prosecution was asking for sentences of twenty years' imprisonment. The judge, I was told, was not unaccustomed to meting out such sentences. There had been a train crash in Zagreb some years earlier and this judge had sent those found to have been responsible to prison for fifteen years.

On the first day of the trial at the adjournment, Drasko took me up to introduce me to the judge. The judge expressed interest in the fact that I was from England and I asked if we could meet to discuss the English judicial system. This started the relationship which went on throughout the trial, and from time to time he would come to the hotel in the evenings and have a drink. I strictly avoided any discussion of the trial in hand. Our discussions were all through an interpreter – either Drasko or some other friend – and it became clear that the judge really did not have much idea about aviation. He had not – and this was half way through the trial – ever been on a flight deck of a civil airliner; he had not seen a high altitude en route chart. It was extraordinary to me that the man should be so bereft of knowledge (this probably was not his fault, I blame the parties) that he was half way through a case without knowing the basic ingredients, and here he was being asked to dish out twenty year jail sentences. But, eventually, the subject of possible penalties came up, and I indicated that I just could not see how these sort of sentences could even be mentioned, or even thought of. Disciplinary action maybe, loss of jobs, loss of privileges, demotion, but *jail sentences* for what was at best human error, and at worst gross negligence? *twenty years* in jail? – what kind of controllers would we have in the future?

Finally, I got around to asking the question 'where did the notion of twenty years come from?' and he answered me as calmly as if I had asked him the time. He said 'that is only forty-one days for each of the 176 people who lost their lives'. Unusually for me I found I did not have an answer, but I finally came up with one in the form of another question. 'What would you do if the planes, instead of carrying 176 people had been Jumbos, like Tenerife a month earlier, and 500 people had been killed?' He waved the question aside. 'Twenty years is the maximum for this offence', he said. I think that that was the moment at which I decided that I disagreed with this process. I disagreed with the law taking this course; it did not fit in with my understanding of what the law is for. It is not meant to be used as a weapon in that sense. The seed had been sown, but I did not really know how to go about what I wanted to do. After all, I was a prosecutor and a prosecutor cannot defend the people he is prosecuting.

Shortly before the end of the trial there was another such meeting with the judge, who indicated that the State Prosecutor would shortly begin his closing speech. He then turned to me and said 'you need not make a speech, but it would be helpful if you would just get up and endorse whatever the State Prosecutor says'. And there it was on a plate for me. I said that I could not do that because I did not agree with the State Prosecutor's position in the case, and he said 'well then, just keep it short' and then I asked Drasko whether there was any restriction on what I could say. He wanted to know what did I mean by that? So I said to him: 'am I correct in thinking that as a co-prosecutor I have the right to make a speech in court?' and he answered in the affirmative. I then said: 'is there any restriction on how long the speech is?' and he said 'no', although he was obviously somewhat bewildered by the question.

I then went up to my hotel room and started to put down on paper what I felt about the whole situation, and that paper finally became my address to the court. It was quite clear to me what was going to happen. It was to be translated into Croatian and read to the court by Drasko.

Drasko saw the first draft a day or two later; he was unbelieving and astonished. He said 'you cannot really expect me to get up and read this. This is a speech for the defence: I have to live here after you have gone, and I have to practise here'. I said 'as far as I am concerned you can get up and you can preface your remarks by saying that you are simply reading this in the vernacular on my behalf'. No, he said. He could not do it.

'You can repeat that refusal', I told him then, 'but you have to do it within the next sixty seconds, so that I know where I stand and I can get myself another lawyer who will do it.'

I think the speech that I delivered in Zagreb then really takes up the story. It says what I felt at the time and indeed what I still feel. The speech satisfied what I felt was my obligation as a lawyer and what I felt was my obligation as a representative of Ruth and her family and indeed, of the other people who died in the crash. To this day I have never met Tasic, although I met some of the other controllers after the trial, but of course Tasic was taken back to jail. I never met any of the other families of people who were killed in the crash, but I did feel that to put controllers away in prison did not help any of the families, did not help to improve the system, and indeed, did not really serve any purpose at all. I said in a television interview, when I was asked for my views, that if I were to return to Denmark and in reply to the question from Ruth's family as to what had happened, that I had to say all the controllers got twenty years in jail that really would not help them at all; nothing at this stage could bring Ruth back or any of the other people. But at least it would be something to think that they did not die in vain. If, as a result of the trial, we could say that the system was being overhauled, that things were being made better and safer for people in the future, that in itself would have been a sufficient monument to those who had died.

But we both know the rest of the story. Let's get on with making it common knowledge.

Richard

Chapter Twenty-two

Whatever had gone before could only have led to the second and final confrontation between Tasic and Pelin, supported in his turn by Bojan Erjavec. It was, after all, on this fourth day that the conflict would centre on the accident itself and on those most immediately concerned with the aircraft involved.

It was not that the evidence given earlier had lacked its dramatic undertones: yet the questions now to be asked no longer posed the issues of other men's capacity to maintain discipline, or a tardy arrival at the console. Grave as these matters were as possible contributions to the collision they must be thrust into the background by the threatening nature of this latest examination.

Erjavec, Bojan . . .

'You were Controller on the middle sector?'

'From 10.00 a.m., yes. Pelin was my assistant.'

'And what happened that morning, when you dealt with JP550?'

'Inex-Adria, yes. He came from the lower sector into ours – into the middle sector: he called in to the sector at 11.00 a.m.'[1]

'And . . .?'

'He reported himself passing FL225 climbing 260. OK, I said; I gave him a Squawk and gave him instructions: fly towards Kostajnica, Zagreb and Graz –'

'And then . . .?'

'And then', said Erjavec, 'I dealt with other aircraft . . .'

'Until . . .?'

'Until, maybe two minutes later JP550, Adria, came back again to say OK, he was level 260, could he go higher?'

'So you said yes? You sent him to 350?'

'There wasn't another higher level available. We had already aircraft in the next levels; so I asked him, can you go to 350 and he said yes, that's all right, I'll go to 350 –'

'Into the upper sector?'

'Yes, into the upper. But first, I said I'll call you back. Then I called Tasic.'

'You spoke to Tasic?'

'No, I didn't speak', Erjavec stated: 'I signalled with my hand that I wanted to talk to him, but he waved *his* hand to say he was busy –'

'And was he busy?'

---

[1] JP550's first contact with the middle sector was at 11.0321″. See page 41.

'I couldn't see from where I was', Erjavec said: 'so I sent Pelin over to him to co-ordinate the climb. He took the flight slip for JP550 with him, to show it to Tasic.'

'Did you discuss it with Pelin before you sent him?'

'It wasn't necessary', Erjavec said. 'He has a higher qualification than I have ... but I saw him talking to Tasic and I heard Tasic say ... "it can ...". Then Pelin told him to let the plane climb: he showed Tasic the target of JP550 on the radar screen while he was there –'

'You say he showed Tasic the target of JP550 on the radar? Was it identified, this target, with code and altitude?'

Carefully, Erjavec said, 'no: it was just the blip –'

'Because you had instructed JP550 to switch off his transponder? You told him to Squawk STAND BY?'

'When he went to the upper sector frequency, yes. If he hadn't gone to STAND BY he'd have been transmitting the wrong code for the upper sector. His code was for the middle sector only.'

'So Pelin showed Tasic a blip. But you were satisfied when Pelin came back to his place? You were sure that Tasic had accepted JP550?'

Erjavec was confident of this. 'Yes,' he said: 'Pelin came back to the middle sector after he'd given Tasic the flight slip and showed him the aircraft on the radar. Tasic hadn't refused the transfer – Pelin would have told me that.'

'So you cleared JP550 to flight level 350?'

'Yes, I did, yes: then Pelin cleared it on the phone with Bec (Vienna) and afterwards I called the aircraft and told him to report to me when he crossed levels 290 and 310.'

'Why did JP550 have to climb? Why didn't he remain at level 260?'

'I don't know', said Erjavec. 'I can't think of any positive reason why he didn't stay there. He asked me for higher so ...'

It is to be assumed that the co-ordination described by Erjavec took place between 11.06 14″ when JP550 acknowledged the controller's promise to 'call him back' and 11.07 40″ when he issued the clearance for flight level 350. The puzzling discrepancies on this score evident in both the Accident Report and in the final Judgement owe much to the testimony of the next witness, and also to the curious history of the flight slip which is to be found in these records.

There is, firstly, the version given in the Report wherein JP550 climbs from the Lower East Sector into the Middle Sector.

| 11.02 44″ | JP550 | Zagreb, Adria 550 crossing 220 |
| | Zagreb | Zagreb 135.8 Good-day. |

'The co-ordination and transfer of the Inex Adria aircraft JP550 was effected in good time. This transfer was carried out on the basis of mutual agreement between the controllers and the passing of a slip bearing flight information ...'

This study has already noted that Zagreb ACC regulations stipulated that flight progress slips should be prepared in advance for all sectors through which aircraft intend to pass. This stipulation appears in the Report along with the information that no such strip was prepared in respect of JP550's passage into the middle sector: and the assertion that, notwithstanding, air safety was not jeopardised since the co-ordination, as stated above, was effected ... 'in good time' and as the aircraft passed Flight Level 220: i.e. while JP550 was still in the Lower Sector airspace ...

Which is a somewhat less than firm indication that the non-existent flight slip ('not prepared for the middle sector') had come into being and had been handed over by the lower sector controller to the middle sector controller 'at the moment of crossing 220: i.e., at 11.0244".'

It is an elusive item, this slip, and yet to reach Tasic on the upper sector. According to Erjavec, Pelin took the slip with him when he went over to Tasic to co-ordinate the climb – at 11.0614" – and again, according to Erjavec, and recorded in the Judgement, he 'pulled the slip out of the rack a little way for Pelin to take'. It is a circumstantial detail, even an impressive triviality.

There could, however, be another interpretation culled from the record of events on the upper sector.

The Accident Report refers to a further departure from ATC standing instructions in that neither had a flight progress slip been prepared for the entry of JP550 into the upper sector. This fact, it is stated, was responsible for that Sector's (controller) 'overlooking' earlier information concerning this flight and its untimely pressures on planning and safety.

It is also remarked that the task of preparing the necessary flight progress slip should have been undertaken by the upper sector assistant controller after JP550 had been cleared to climb to Flight Level 350: clearly that could not have been done in advance since the original flight plan for the aircraft had not called for altitudes in the upper sector: and equally clearly, it could not subsequently have been done, as required, by the assistant controller since he had not been at his post.

The onus was therefore on Tasic: but *he* had not been able to prepare the slip because of his workload ...

So that Tasic saw no slip concerning JP550 until, it is claimed, Pelin approached him 'taking the flight slip for JP550 with him' ... at 11.0614" or 11.07, and that co-ordination was completed by 11.0740".

There is an implicit assumption in the Judgement in that phrase 'the Court finds determined that Pelin took a strip from the sector console and carried it to Tasic ... for ... co-ordination', since, although there is no direct reference to the *time* at which this alleged action took place, the statement is subsequently given some apparent credibility thus: '... co-ordination between Tasic and Pelin took time [sic] between 11.07 until 11.0740". *Forty seconds is more than enough for a controller with Tasic's licence, (to) determine and (make) the right decision about accepting or rejecting* the co-ordination and giving clearance to the plane to enter his sector.'

That may be: but in view of the damaging weight thrown against Tasic by this pronouncement the actual time of the transfer of the slip into Tasic's hands is treated with a remarkable flexibility since the actual time reference in the Accident Report alludes *not* to the transfer of the slip, but instead, to the instruction given to JP550 at 11.12 12″ that the pilot should change to the upper sector radio frequency.

The Report then goes on to say that it was immediately prior to, or perhaps at that time, that the flight progress slip which had been used in the middle sector was passed into the hands of the upper sector controller. Similarly, although in the Judgement 'the Court finds determined that Pelin took a slip from the sector console and carried it to Tasic . . . because of (for) co-ordination', yet *the only time firmly linked to this act* is quoted in a later page thus: 'Finally, it is confirmed through Tasic accepting the strip at 11.12 12″ where it is evident that the plane is ascending to FL350. Yet Tasic does not react.'

Forty seconds, said the Judgement, is more than enough time for an experienced controller such as Tasic to appreciate a situation and make the right decision: but if the hiatus in the evidence noted above is examined, namely, that the handing over of the flight slip at 11.07 is in the one case merely *implied* by the linked statements 'co-ordination took place between 11.07 (until) 11.07 40″ and 'forty seconds is more than enough time etc. etc.' and in the other, *expressly stated* to have taken place at 11.12 12‴, then there is certainly room for reservations on the question of Tasic's forewarning. It might be timely, now, to look again at what Tasic was doing at this stage.

11.11 42″	Zagreb	(to BE778) Roger . . . if you need descent before Split, report it –
	BE778	Yes . . . descent before Split
	Zagreb	OK
11.11 53″	Zagreb	Finnair 1673, go ahead now, copy 1673 go ahead –
	F1673	Finnair 1673 passed Graz at 10, level 390, estimate . . . break . . . 1
11.12 10″	Zagreb	Finnair 1673, report passing Delta Oscar Lima, maintain level 390, Squawk Alpha 2310
11.12 20″	F1673	Will report passing Dolsko at 390
11.12 24″	LH310	LH310, Sarajevo at 09, 330 Kumanovo 31
	Zagreb	LH310, contact Beograd, 134.45, sorry 133.45, good-day
	LH310	Good-day
11.12 38″	Zagreb	Good-day
11.12 40″	OA172	Zagreb, Olympic 172 –

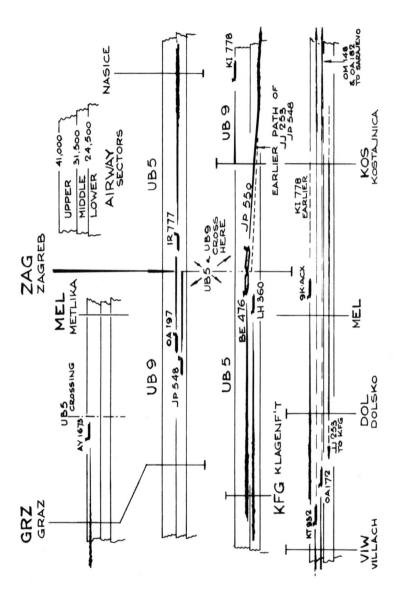

Fig. 4 Zagreb air traffic situation frozen at 11.1438″. Flight levels.

112

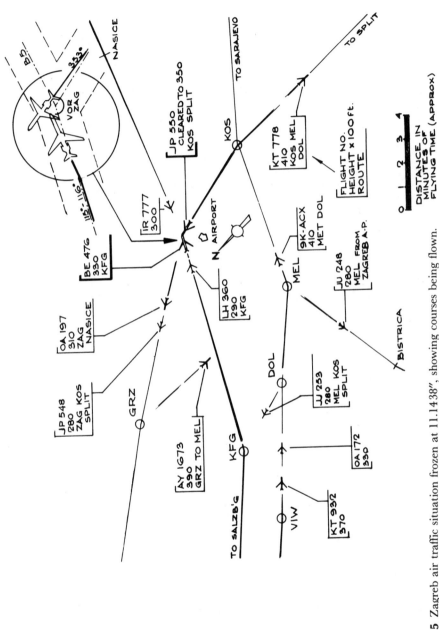

Fig. 5 Zagreb air traffic situation frozen at 11.1438″, showing courses being flown.

113

It is claimed that Tasic received the slip at 11.12 12″, and there is, indeed, a 14-second gap between the commencement of his transmissions to Finnair 1673 at 11.12 10″ and LH310 at 11.12 24″.[2] Both aircraft called Tasic during that interval: but he could, at the same time, have been appraising the slip. It would by no means be for that leisurely 'forty seconds' since he was thereafter, as the control tower tapes have shown, continuously engaged with the overflying aircraft OA172 and BE932. And if Tasic *did* receive the slip at 11.12 12″ he had one minute and 52 seconds – eroded as noted below – before JP550 made its first call to the upper sector: . . . at 11.14 04″, exactly, 34 seconds before the collision. The traffic situation at that time is shown in Figures 4 and 5.

[2]The '14-second gap' recorded above should not be taken too literally. The actual transmission to Finnair 1673 at 11.12 10″ took *at least nine seconds*. Tasic would have finished his call, therefore at 11.12 19″ or even at 11.12 22″.

Chapter Twenty-three

Pelin, Gradimir ...

'Yes, I was on the morning shift', Pelin said: 'I went on duty at seven o'clock with the others. I worked as assistant controller on the middle section from 11.00 with Erjavec.'

'He used the radar?'

'Erjavec? Well, no: he isn't qualified for that. The radar was only being used for reference: he used procedural control.'

'And until JP550 asked for a higher level there were no problems? You were both familiar with all the traffic in your sector?'

Pelin confirmed that this was so. No, they had had no problems . . . only when JP550 wanted to climb . . .

'Then? . . .'

'Then Erjavec tried to attract Tasic's attention: he wanted to clear the aircraft into Tasic's sector –'

'He spoke to Tasic?'

'No – he signalled by hand that he wanted to talk to him.'

'And?'

'And ... Tasic waved *his* hand – he didn't want to talk just then ...'[1]

'Because he was busy?'

'He was working some aircraft, yes,' replied Pelin. It was not quite assent.

'And so ...?'

'So Erjavec told me to go to Tasic to co-ordinate. That was at 11.07 – he gave me the flight slip and I took it to show Tasic.'

'Tasic was alone on his sector?'

'Yes, he was alone – he was talking to the aircraft. I waited until he'd finished, then I asked him about JP550.'

'What did you say to him?'

'I asked him: "Can the DC9 take FL350, Zagreb 11.15,[2] Graz next?" And I handed him the slip to look at.'

'And Tasic looked at it?'

---

[1]Between 11.05 44″ and 11.06 59″ when the alleged exchange of hand signals took place Tasic had been working IR777, Monarch 148 and Grumman 9KACX. See page 46 for transcript.

[2]There appears to have been a strange failure to attach significance to this time of 11.15 which would have meant one minute's separation between BE476 and JP550. This estimated time of arrival over Zagreb appears, as quoted, in the Judgement's review of Pelin's evidence. None of the relevant tapes of the control tower radio and telephone conversations mentions this E.T.A. Zagreb. Nor does the report. Tasic first heard the E.T.A. from JP550 Zagreb at One Four at 11.14 10″, 28 seconds before the collision. See page 53.

'Yes', said Pelin flatly: 'He looked at the slip. After he'd looked he said, where is it? Where's the plane now? So I showed him the target on the radar.'

'You pointed with your finger? You put your finger on the target?'

'Yes', said Pelin. 'I knew it was JP550 because it was identified with a Squawk code on the middle sector radar. And then Tasic said, "Yes, it may (climb)". But then –'

'Yes?'

'But then I saw the target of the plane coming from the direction of Metlika[3] so I said: what about him? And Tasic scratched his head and said: "OK, when they cross . . ." I went back to the middle sector to check that I'd pointed at the right blip for JP550 and then I came back to Tasic. We looked at the screen, together and when they crossed, Tasic said it again . . . "It may". So I turned to Erjavec and told him:[4] yes, climb it. I'm sure Tasic must have heard me, I was standing right next to him: much closer than to Erjavec.'

'So Erjavec then gave JP550 permission to climb?'

'Yes. And I went back to my own place and called Vienna on the phone to clear the aircraft there.'

'That was at what time?'

'You've got that', said Pelin: 'it was 11.07 50″ when I called Bec. Next I heard JP550 report crossing 290, then Erjavec telling him to call crossing 310 and giving him the upper sector radio frequency. Tasic turned towards me and took that slip. After that we worked normally until . . .'

'Yes? Until?'

'Until I heard the Lufthansa pilot talking about the crash. I went to find out what was happening – what had happened: first I went to the lower sector, and then the upper – Tepes was there.'

'Where was Tepes?'

'Sitting between the procedural control position and the radar with Tasic next to him. I looked at the radar for Adria JP550 because I knew it was close to Zagreb. I couldn't see it, so I asked Tasic: Where is he?'

'And Tasic, what did he say?'

'He said: "WHY?"'

A silence. Then Pelin said: 'I looked at the slips in the console and I saw that the British plane should be over Zagreb at One Four, the same time as Adria. I told him to call Adria and watch the radar screen. When Adria didn't reply, I said call the Trident.' Pelin's hands made a small gesture of hopelessness. 'There was nothing', he said: 'Nothing on the radar. And nobody replied.'

---

[3]As stated in the Judgment. In contrast to this wording the Accident Report says that: The upper sector controller stated that he remembers that (Pelin) showed him *some* aircraft in the vicinity of *Kostajnica*: according to Pelin this refers to Olympic 182 and 330 and OM148 at level 370. Tasic, in his own version of this episode says that, after Pelin asked him for permission to climb JP550 he 'checked the slips on his own console and noticed that he had OA182 near Kostajnica at 330 and BE476 at 330 approaching Zagreb'.

[4]Or signalled, states the minutes of this evidence.

'Let us come back to the flight slip. It was correct, clearly marked?'

'It was correct, of course, with all the necessary data.'

'Data on the altitude of JP550?'

'Yes: the plane was climbing for FL350 – 310 was crossed out on the slip to show he'd left that level. If he was still at 310 we'd have put a circle round that figure, not crossed it out.'

'And the target you showed to Tasic? You were sure it was JP550?'

'Of course I'm sure. Certainly!' Pelin clearly felt that he had already answered this question: 'It's just as I said. I checked on the middle sector radar: he was identified with a Squawk.'

'And you gave the flight slip to Tasic and he looked at it, you say?'

'Yes'.

'Why didn't you mention carrying the slip with you when you gave evidence to the Court of Enquiry. You didn't mention the slip at all in your earlier statement. Why was that?'

'I didn't know', said Pelin: 'I didn't know that it would be necessary to go into all this detail. But the evidence now is the same evidence I gave to the Court of Enquiry ... maybe they didn't write everything down. I don't know.'

'Well, you went over to Tasic. Do you know whether he was working procedurally or by radar?'

'I can't say', said Pelin: 'I don't know that, either.'

Those who looked on had witnessed an extraordinary interweaving of statement and counter-statement. They had also been the spectators of an equally novel attempt to sound for truth by confrontation: as if, perhaps, the most strongly voiced version of events and the most ironclad demeanour of innocence were of themselves trustworthy, if not decisive, indications of integrity. It would transpire that this form of test was indeed to be so regarded. On occasion. And never in Tasic's favour.

The fuse for the second of these scenes was now touched off by Judge Zmajevic. The eruption which followed left no question in the minds of the observers as to which man had been destroyed.

'We should remind ourselves of Tasic's evidence', said the Judge: 'that he told Pelin that if the plane makes 310 before Kostajnica, it can climb to 350. Or if the plane couldn't make 310, then it could only climb to 350 after Zagreb.'

The Judge beckoned to Tasic to stand beside Pelin. 'Now', he said: 'Pelin claims that you did not tell him that and that it was impossible in any case because you had Olympic 182 near Kostajnica at level 330. What do you say to this?'

'That he lies', said Tasic, hotly. 'I *did* tell him and I was never shown the strip when he came to me: I didn't get it until 11.13, maybe a minute, or a minute and a half before the crash –'

Pelin interrupted at once: 'You gave permission for the climb', he said, emphasising each word: 'you were holding the slip in your hand. I came over to

117

discuss the ascent with you. Be sensible –'

'You lie!' said Tasic, furiously. 'You lie! I tell you to your face – you lie!'

Both men were shaking. The force of that encounter stilled the room, then triggered the journalists into a frenzy of scribbling.

'It's true', Pelin was saying: 'It's true: and I warned you about the aircraft near Kostajnica – I found out from Tepes, it was Olympic 182 –'

Pelin began to repeat his defence statements. The words fell on ears still ringing with Tasic's bitter rejoinder.

It seemed that there was little more to be mined from Pelin's evidence: but once more the questioning touched on why JP550 had needed to climb at all.

'Tasic says you insisted on climbing the DC9: couldn't he have remained at 260? Couldn't he have flown through Zagreb District Control at that level?'

'As far as I know, yes. There was no question of insisting – that isn't right. He could have stayed at 260 and gone through our sector like that. He just wanted to go higher: I don't know why', said Pelin: 'he just preferred a higher altitude; we didn't have one so we had to talk to Tasic.'

'To talk to him? Is that better when you are making a co-ordination? Better than using the intercom?'

'Better?' Pelin frowned: 'Tasic was sitting there, right next to us. It was natural to talk to him directly.'

'And you could hear each other distinctly? There was no possibility of confusion?'

'There was no confusion', said Pelin. 'I know what he told me and what he didn't tell me. He said "wait till they cross, then OK, it can climb".'

There was time, before he stood down, for Pelin to deliver another formidable thrust: 'It wasn't my job to make decisions anyway', he said: 'I was the assistant controller. I can't do anything without the controller's permission. And *that's* my whole defence.'

It was more than enough, since it enabled Pelin, like his colleagues, to distance himself from the contagion of a burden shared with Tasic and shelter in the lee of that wall which now existed between them.

The foundations for that wall had been laid long before, on the occasion of Judge Jakovac's visit to the crash site: it was, he had said, 'a very serious matter when you find men who are directly responsible for a disaster of this magnitude'.[5]

Delic, Munjas and Dajcic, themselves threatened, had been at pains to defend their own conduct: at the very least, to avoid condemnation. Yet, creditably enough, they had not sought to deflect responsibility and Delic, indeed, had offered Tasic no small comfort. The wall would rise without them, course on course, provided by Hochberger ... 'we weren't very busy: we chatted a lot ...'; by Tepes ... 'I asked him if there was a lot of traffic and he

[5]See page 68.

said: "no, not a lot . . ." '; by Erjavec who had, he claimed, initiated a proper co-ordination by sending his assistant Pelin to Tasic 'with the flight slip'; and by Pelin . . . 'Tasic looked at the slip . . . he didn't tell me what he claims he told me . . .'

But Tasic, too, had inadvertently reinforced the safer side of that wall, for his angry challenge, seeming to injure not Pelin but himself, had also offered his judge an apparently simple choice. 'You were holding the slip in your hand', said Pelin: 'be sensible . . .'

It sounded very much like the truth: much more so than the only words which Tasic could find in answer:

'You lie! You lie! I tell you to your face -- you lie!'

Chapter Twenty-four

Presumably, everyone is lying wrote Olaf Ihlau, covering the trial for the *Süddeutsche Zeitung*;

> and it seems doubtful whether the appearance of fifty witnesses and experts during the next two weeks will really clarify the decisive issues or whether, for example, Tasic was overloaded, whether the other air traffic controllers had transferred the DC9 properly to his control and with his consent and whether permanent lack of discipline and infringement of the supervisory duty in Zagreb Control Centre did not inevitably lead to the disaster.[1]

The very title of Ihlau's report – 'A Scapegoat for the Eye of a Needle' – reflected that foreboding of Weston's so early in the day: that it was emerging now as an inescapable consensus among observers at the trial spoke volumes for the orientation so far – and so obviously – given to the proceedings.

> Gradimir Tasic, as has become clear in only a few days in Court, has been psychologically isolated in this Zagreb trial and been marked out as the scapegoat. The other seven air traffic controllers, who have all incurred previous disciplinary punishment, are trying to push their guilt onto an unpopular colleague who is threatened with up to twenty years in gaol.[2]

Without surprise readers noted the distinctly angry tone of this piece. While German anguish would in any event have brooked neither evasion nor excuses, the transparent readiness to offer Tasic as a sacrifice had evoked a withering response embracing the alleged maleficence, demonstrable ineptitude and universal inadequacy of Yugoslavia's air traffic control system: and, by association, its administration.

Ihlau had by now heard the evidence of Tomas Samardzic, Deputy Director of the Yugoslav Federal Civil Aviation Administration and had drawn his own conclusions from Samardzic's assertion at the witness stand that 'radar is not all-powerful'.

'That may be so', commented Ihlau testily, but nevertheless, the

---

[1]'Ein Sündenbock fürs Nadelöhr' in *Süddeutsche Zeitung* 21 April 1977.
[2]*Ibid.*

statement sounded a little like an excuse. Because Samardzic had to admit to the Court that the Yugoslav air traffic control had had Swedish radar equipment . . . 'as an additional source of information' at their disposal, and that he himself had ordered their operative use in January, 1974.'

'But because of missing auxiliary equipment the radar screens in the control towers had not been expertly adjusted and consequently they were of very dubious assistance.

Inescapable, too, now was the conclusion that it was by no means sufficient that Tasic should be made to pay for these deficiences: as if his obliteration would also wipe away the outrageous revelation of an admittedly threadbare provision for that immense air traffic, as well as the shoddy expedients and anarchism which characterised it.

It was not at all sufficient: but given the manner and terms of the indictment and the drift of events in the courtroom, the possibility of a rough, pragmatic and totally irrelevant justice for Tasic seemed extremely likely. There was, therefore, nothing else for impassioned, but frustrated, journalists such as Olaf Ihlau, except to plant barbs in an attempt to puncture the complacency they suspected.

This is certainly not the only example of technical backwardness, sloven-liness, inadequate safety measures and lack of discipline in the Yugoslav air traffic control system which has so far been brought to light in the Zagreb trial. But it still remains to be proven that there is a cogent chain of causation linking such negligence with . . . the disaster . . .

It is known that the British members of the joint enquiry board which had been set up after the collision... were horrified by the chaotic conditions in the Zagreb air traffic control centre and the 'hasardspiel' – the 'game of chance' practised there . . .

Ihlau appeared to temper the piece with a grudging modulation. '. . . neverthe-less, western pilots flying regularly to Yugoslavia (were) prepared to offer Zagreb the categories "satisfactory" to "good".' He also quoted the observation by Ante Delic that 'more than 700 000 aircraft, excluding military planes, have been channelled through this aerial "eye of a needle" over the Balkan peninsula'.[3]

Yet that information did nothing to explain the fate of the two aircraft which had *not* been channelled through the needle's eye. It was quite impossible, too, not to ponder the implications of the claim – also made by Delic – that 'despite the difficulties of staffing and technology, conflict situations were . . . relatively rare: no more than "three to five" near misses annually'.[4]

Any comfort intended by this ratio was quickly dispelled by cold logic: two

[3] *Ibid.*
[4] These, of course, are 'reported' near-misses. The actual number of such occurrences is shrouded by reluctance as well as failure to report many incidents.

errant light aircraft, for instance, counted as one 'near miss' and jeopardised perhaps two, three or four lives: but if one moved up the scale of probability – since the figures for 'near misses' did not specify aircraft types – and the two light aircraft became Tridents and DC9s, then it was very possible that each occasion put at risk *another* 176 lives or even more. And at the end of the scale, of course, was the possibility of such an occurrence involving two really large aircraft – B747s, L1011s or DC10s, each with their two or three or even four hundreds of passengers.

It may well have been true that the British investigators had been horrified in the face of these possibilities: but they had, in fact, maintained a scrupulous professional discretion in the matter of utterances for publication and whatever had passed between them in private had certainly not been blazoned abroad.

Yet if witch-hunting was no part of their brief, that idea inhibited no one else: Ihlau's darts, for example, were cast in such profusion as to strike anyone within range and it was unfortunate that Tasic stood nearest in the line of fire.

Tasic's version is disputed by the other air traffic controllers. They maintain that Tasic had always wanted to do everything by himself and had not complained about being overloaded. But the tape recordings of the radio communication during the fatal last five minutes ... reveal that the principal accused was maintaining constant verbal contact with 11 aircraft and, in addition, carrying on four telephone conversations with Belgrade. At the same time he reacted calmly to this and appeared confident. *Possibly, he hoped – if he had not perhaps even temporarily 'forgotten' the Trident 3 – that somehow it would turn out all right as with earlier precarious situations, and he would be able to cover up the whole thing afterwards.*[5]

Yet Tasic was not the primary target and in acknowledgement of that fact, Ihlau ended:

> ... as regards the compensation claims put forward by the bereaved dependents, there is finally one question of special importance, and this is how far the Yugoslav captain was also responsible. Nevertheless, the admission that it was not the individual failure of one air traffic controller which led to the death of 176 people, but that there were probably also a few other determining factors, can be deduced indirectly from the changes which have been made in the Yugoslav air traffic control system in the meantime. The chief air traffic controller, Civikovic [*sic*] let it slip in his testimony that the technical equipment and the division of the flight zones had been rearranged . Also, eight additional controllers had been moved from Belgrade to Zagreb and the time interval for aircraft flying at the same time altitude had been doubled from ten to twenty minutes.

This unconcealed contempt for such stopgap repairs reflected a deeper fear and

[5]Authors' italics.

resentment of those who for so long had sheltered behind the mystique of professional aviation: and who had unquestionably enjoyed the kudos associated with their status. So rudely displaced from that eminence, these men were now seen to be not only intellectually earthbound but, in addition, to be as vulnerable to moral weakness as any other being.

It was not an opportunity to be ignored by Der Spiegel and it was in the course of a bitter and provocative post-trial review that this journal, among other achievements, offered an 'explanation for one hitherto totally unanswered' question.[6]

At the beginning of December the accident report was already completed, yet it was not immediately published as first promised, but submitted to Belgrade so that the government would be in a position 'to comment on it'. The government found that for reasons of 'higher interest' the Yugoslav Aviation Administration could be released from its obligation to publish the details.

The criminal proceedings in Zagreb were also

... held under the banner of 'higher interest' and 'socially valuable information' – as the judges put it. Even so, the bill of indictment brought by the public prosecutor, Slobodan Tatarac, vaguely mentioned that '... apart from acts of negligence on the part of the airline pilots, there were other contributory factors which led to the collision of the two aircraft'.

A great deal was made of the acts of negligence, even of the wretched housing conditions of the principal accused, Tasic: but nothing was said about the other factors [and] the key question remained totally unanswered: why did Erjavec, the controller of the intermediate air corridor, insist so obstinately that the DC9 should transfer to the upper corridor, which was already too busy for Tasic?

It was doubtless a cruel and dramatic interpretation of the procedure which Erjavec had decided to follow, but here again, the tone was in keeping with public cynicism and frustration. Too many obvious gaps in the record appeared to have been tacitly accepted by the Court: Der Spiegel therefore gripped this particular bone in its teeth and worried it with ferocity.

...towards the end of the trial, the Zagreb newspaper 'Vus' was certain that '... some important matters are still unknown' while the rival Belgrade publication 'Nin' was more explicit: Erjavec's insistence on directing the DC9 with its German tourists on board towards a collision course with the Trident remained 'inexplicable'.[7]

And not only to these two papers: but in supporting this assertion Der Spiegel,

[6]Der Spiegel No. 22/1977.
[7]Der Spiegel No. 22/1977.

mindful of a national indignation, did not fail to include a vitriolic aside on the quality of the Yugoslav airspace administration:

The counsel for the defence of the principal accused, the eminent Belgrade lawyer Toma Fila, wanted to clarify this too, and on 4th May brought forward some high-ranking officers of the Yugoslav armed forces as witnesses for the defence . . .

Lieutenant Colonel Sava Zivkovic, responsible for overall safety within the Yugoslav airspace had been called to the witness box earlier but his evidence, according to 'Nin', '. . . more than destroyed him', because he declared that he was 'not familiar' with the details of the air safety regulations, that he did not know the instructions regarding the use of radar equipment and that he was 'not at all' acquainted with 'cases endangering safety in 1976'.

It was time, at last, for the promised revelation:

The public was excluded from the court while the officers were examined, for reasons of 'security'. And yet the secret which the district judge, Branko Zmajevic, had avoided revealing for a month, was still leaked to the public.

The intermediate air corridor in which the DC9 could logically have continued its flight to Cologne unhindered, had to be hastily cleared by air traffic controller Erjavec: the corridor was barred at the time for a special flight of the presidential aircraft belonging to the Yugoslav head of state and party leader, Tito.

Two days previously, Tito had received his Roumanian state visitor, Nicolae Ceaucescu, at the Slovenian castle of Kranj, had gone hunting with him, and had flown back with him over Zagreb to Belgrade on the morning of that black 10th September. Tito's Boeing had already landed at the Belgrade airport of Surcin at 10 a.m., but the military air traffic controllers were overconcerned about the President's safety and had still not declared the corridor free. It was still off-limits for civil tourists.[8]

And it was clear, summed up *Der Spiegel*, that it was Gradimir Tasic who would pay for the incompetence and confusion of the Yugoslav airspace administration.

[8]*Der Spiegel* No. 22/1977.

Chapter Twenty-five

To Ronald Hurst

From Richard Weston
London,

August 26th

Dear Ronnie,

Following our recent telephone conversations I am now going to have a crack at the follow up part to my last long letter, i.e. covering the events as I recall them immediately following the crash ... In about an hour's time I am going down to Leytonstone to get some more papers out of store. These will reveal a lot more names and letters, which I cannot recall offhand and, of course, they will help to fill in some other details later: but right now it is 6 o'clock in the morning and I feel like getting down to this. It will necessarily be disjointed, and some of the things will be out of order, but at least if I get it on to paper we can straighten it out in due course:

Ruth came from a very small village on the northern coast of Jutland called Voersaa. Her mother was named Else and her father Villiam. Her father was a fisherman who used to go out at 4 o'clock in the morning, keep for the family whatever fish they wanted and then take the rest of his catch to market: hence, whenever we used to go out to dinner, Ruth was not a great one for ordering fish. Her mother looked after children at the local kindergarten; her elder sister was married and had a little boy called Thomas.

Her elder sister's name was Inge Merete. Her younger sister, Elin, bore a striking resemblance to Ruth. During the course of our relationship Ruth used to joke that because she spoke Scandinavian languages fluently, British Airways never put her on a flight to Copenhagen, but of course, she had the usual travel facilities afforded to airline employees, and she used to make fairly regular trips home. She had often asked me if I would make the trip with her but somehow I never got around to it: but she did make me promise that one day I would take her home and meet her parents. Mercifully, there was no way to know how faithfully I would in due course keep that promise.

Needless to say, she was the apple of her parents' eye. They were a simple family who lived a simple life. The other two daughters lived and worked nearby; the elder in a furniture store and the younger in a bank. But Ruth was the one who had gone off into the world, who was sophisticated, elegantly uniformed and qualified. She was the one who spoke foreign languages, appeared on television and in brochures, etc., etc. The only time when I had spoken to either of her parents was when Ruth was visiting home and would telephone to speak to her. It turned out that they both understood enough English for me to be able to communicate with them but on the telephone the conversation was always short: 'May I please speak to Ruth', I would say, and she would be on the line.

Reverting to the evening of September 10th; I suddenly found myself in the middle of a personal disaster, involved in a professional situation because aviation was a large part of my own professional occupation, and in the middle of what was suddenly the worst mid-air collision in the history of aviation. On the evening of Friday, September 10th the suddenness and the finality of it was difficult to comprehend. Saturday and Sunday seemed almost like normal days. There was no point in going to Yugoslavia; what for? To look at the wreckage? To look at the bodies? I really do not think that the meaning of what had happened had really sunk in but clearly I had to communicate with her parents. But I was also a coward and I did not want to tell them. I finally spoke to Frank Carver, the British Airways agent in Copenhagen; he told me that he was flying to Aalborg that evening. That is the nearest major airport to Voersaa and he was going to break the news to Ruth's parents. I asked him to ring me when he got there. He was known to them because he had acted as an intermediary from time to time when Ruth was going to Denmark, either on a service which did happen occasionally, or for a trip: he would ring and advise her parents. In many ways, his visit must have been the worst moment of all. To Ruth's parents he was Ruth's boss and when he rang it was always exciting because it was something to do with Ruth and her job, but to turn up there as he did and have to tell them the final news was something for which I think he ought to have been given a medal.

As I said in my last letter, over that weekend British Airways were already starting to organise a pilgrimage to Zagreb for the relatives, for the following Tuesday, and Derek Stafford, Ruth's cabin crew supervisor had invited me to go along. The invitation was also made to Ruth's parents. British Airways offered to fly them to London on the Monday, to put them up in a hotel and fly them down to Zagreb on the Tuesday. They declined; as far as they were concerned they never wanted to go on, or for that matter, to see an aeroplane. I told them that I would go on Tuesday and that I would then go straight on to Denmark. The arrangements were as follows: I was to stay overnight at Heathrow on the Monday and on Tuesday morning I was to fly to Zagreb and on arrival there was to be a non-denominational memorial service, followed by lunch, followed by various other services. As far as the majority of the English relatives were concerned, this was to take place at the English church in Zagreb and then the group was to be brought home. I made arrangements with British Airways to fly straight from Zagreb to Copenhagen, connecting to Aalborg after the afternoon service. Fortunately the SAS schedule fitted in with this programme.

It hardly needs to be said that the flight to Zagreb was, I suppose, the most wretched flight that I have ever been on.[1] I remember reading the morning paper on the way down. There was a report of a near-miss over Paris the day before when a British Airways Trident flying in cloud had deciphered the French message being passed to an Air France plane to pass over the same beacon to which he was cleared, at the same altitude. According to the newspaper the pilot actually felt the wash of the Air France plane as it went past him. By now, as we know, the air traffic controllers had had the blame for the Zagreb accident laid firmly on their shoulders. The details would not emerge for some months, but the basic cause was said to lie with the air traffic controllers and I think that made everybody feel a little better. When God does it all by himself one does not have anybody to throw stones at, but at least now we had a target. The air traffic controllers had all been arrested and, as I said, at that time, like everyone else I fervently hoped that in due course they would all be hung for this murderous piece of negligence.

On arrival in Zagreb in the pouring rain, one could only pity the poor British Ambassador who stood at the bottom of the aeroplane steps shaking hands with everybody as they came off the plane, mumbling something like: 'I am so terribly sorry'. Through Passport Control for those who had passports: some had not got them or had not had time to get them, but the formalities were waived. Two other planes arrived at the same time, one from Cologne – because of course the second aircraft was carrying tourists back to Cologne – and a third one from Istanbul, because there were many Turks on the British Trident. Into the buses outside the airport and then into what seemed like a holy war on the pavement next to the bus behind. The problem was that the Turks did not want to go to a memorial service, they wanted to go to the morgue; they wanted to see the bodies – it was a religious matter. And at that moment I suddenly thought, what about the bodies? It suddenly became terribly important to know whether they had found all the bodies, whether they had identified all the bodies and then it narrowed itself down from bodies, to Ruth. Where was Ruth? Was she comfortable? Was her genuine golden Danish hair all in place? She used to say: 'This is the real McCoy you know, this is not dyed'. She was very particular about her appearance. What about it now?

We went to the service at a building rather like the Festival Hall,[2] new, clinically clean and adequately sombre. The City of Zagreb turned up to pay its respects to the dead and those left behind. There was a sense of shock; a sense of dignity; a sense of finality. There was an orchestra who played the funeral march, there were the flags of the various nations whose nationals had been killed, there were dignitaries and diplomats and the whole thing was organised perfectly: it was a tribute very much on the level of a State occasion in Britain. Don't we always handle these things well? Then on to lunch at the Esplanade, with the powder room becoming the dry your eyes room and the lunch becoming like a wake. I went and sat myself in a corner and was swiftly joined

[1] Yet in that distinction it was shortly to be surpassed.
[2] In London.

127

by a man who introduced himself as the Operations Director of British European Airways. Somewhere Ron, you will be able to find out whether this was before or after the merger if it matters at all.[3] If it was before, then it was British European Airways, and if it was after, it was British Airways European Division . . . they talk about ED for European Division and OD for Overseas Division. You probably know that the merger is probably the biggest non-merger in history, anyway. My recollection is that the merger had been announced. Be that as it may, I was joined by a man who introduced himself as the Director of Operations. I had to ask him about the bodies; and he produced Dr Preston who was the British Airways doctor.

Immediately after the accident I learnt that crash investigation teams had been despatched, together with a team of pathologists from RAF Halton, and British Airways then told me it was not their policy to announce identifications until everybody had been identified for the following reason: 'There are 176 people. It would be grossly unfair if we announced the names of 170 bodies who had been identified and left six people without information.' Of course if you are involved in aviation you know that these crashes do terrible things to the people involved in them. There have been ghastly stories about anguished relatives opening coffins after their loved ones have been returned, only to suffer further horror. I now wanted to know, suddenly, where Ruth was, whether she had been identified and was I going to be able to take her back to Denmark for burial? The procedure, I was told, was as follows: the airline had a contract with Kenyons the London funeral directors: they would fly a Vanguard out with a large number of crude coffins, the Vanguard would then load up all the coffins, fly on to Istanbul, unload the Turkish coffins, fly back to London with the British coffins and Ruth would then be despatched in her box to Copenhagen. What an irony – she did not like flying when she was alive and was about to make a grand tour of Europe in death. I declared that I was not having any of that, and could I please keep the promise which I had so often made and take her back to Denmark myself? Everybody was very nice; they thought arrangements could be made, but I would have to wait, nothing could be done until the identification process was complete. So I asked the next question – has Ruth been identified? They finally told me, privately, that the answer was affirmative – yes, she had been identified. How was she? I asked. They said she was fine; everybody was found sitting in the plane. They had all died very quickly as a result of decompression at 33 000 feet. The plane had crashed almost intact and when they went into the fuselage everybody was found sitting neatly strapped in their seats.

I suppose, in retrospect, that was the kind thing to say, and I should have left it at that, but there was still a big fuss going on – the Turks wanted to go to the morgue, and finally it was explained to me. Things were in a bit of a mess; the morgue was not the place to be at that time. On a previous occasion a group of

[3] No, but for the sake of consistency we have used the current designation, British Airways. (R.H.)

relatives had insisted on going to the morgue; the airline had refused. After lengthy negotiations the airline finally agreed to allow two doctors who had accompanied the relatives to visit the morgue. One of the doctors had a heart attack and the other one had to be helped out. Apparently, the sort of mess that aeroplane crashes make of those directly concerned is not the sort of thing to which the average general practitioner is accustomed. The answer was no, no and no again, you cannot visit the morgue.

Armed with half a promise that I could come back and get Ruth when the Health Authorities (what a misnomer) had finished with her, I set off for Copenhagen. It was a dull trip. I then set off for Aalborg. I was met by Inge Merete and taken to the White House (the principal hotel there); it was quite late by now. The next morning Inge Merete picked me up and took me to her parents' house about 45 kilometres or half an hour away. To say the meeting was difficult would be a masterpiece of understatement. It was painful. But even then, and as I have come to know them better over the years, they were delight-ful people: proud, hospitable, warm, intelligent, and shattered. As one might expect of a Danish fishing village, the whole outfit was spotlessly clean. No, they did not want to go to Zagreb, they did not want to go to London, they did not really want to deal with anything. But yes, please, they would very much like Ruth home; and would I, could I, please deal with everything as I saw fit? After a few hours I left and returned to the UK. There was so much to do. The funeral had to be arranged, but most important I had to get the airline, the Yugoslav authorities and the Danish authorities to agree to let me take Ruth back to Denmark. I cannot remember, without my file, what happened: all I know is that this took another ten days. It seems in retrospect that I literally worked round the clock every day to make the arrangements. I remember specifically arranging that on arrival in Aalborg I would have Ruth's coffin changed; I did not want her parents to see the plain oak box which Kenyons were going to fly out from London as one of the contract batch. At least I wanted her to look decent when her parents saw her. I remember going out to RAF Halton to talk to Group Captain Balfour who was in charge of the gruesome business of sweeping up. He told me that there was a body plot – a body plot is a map of the surrounding area, which shows the site on which the aircraft crashed and where the various bodies were in relation to the aeroplane. It will be recalled that the nose of the Trident was in fact severed by the wing of the DC9 and as the plane came tumbling out of the sky, so the bodies tumbled out of the plane. The two planes crashed several kilometres apart: elsewhere in the book we have noted the approximate closing speed and angle of impact, so that it is not surprising that they fell some way from each other.

From the body plot and my conversations with the Group Captain it appeared that the bodies were in fact spread out over the countryside, and possibly the less investigation I did, the better off I would be. I remember that the amount of work which I did during those ten days was actually so concentrated that I did not really have time to stop and think about what I was

129

doing. It had become almost more important to get Ruth home through this tremendous amount of red tape than it was to sit down and miss her or cry over her. I showed up at British Airways on the appointed morning. I travelled down with Mr Charles Stuart, a director of the airline with whom I have since become very friendly. He was on his way to attend a funeral and memorial service for some of the passengers whose relatives did not want them repatriated. I had made arrangements to arrive in Zagreb to collect the coffin and fly back to Copenhagen where I would switch to another SAS flight there to go on to Aalborg. At Aalborg I was to be met by the funeral directors who were to collect the coffin and change it for a better coffin and place this in the morgue overnight. The funeral procession would leave there the following morning to travel to Voersaa.

On arrival in Zagreb the British Airways representative met me with the SAS representative and I was told that in due course I would be able to board the Copenhagen plane. When this was called I was taken out by car to the plane and alighted by the steps. I indicated that there was no way I was boarding the plane without Ruth, and could I please go round to the baggage door and make sure the coffin was loaded and could I also please examine the paperwork to make sure, insofar as was possible that I was taking the right person to Denmark. Finally, the baggage trolley arrived with the coffin and the papers. The airport staff removed their hats and stood while I examined the papers. I think it was not until that moment standing by the coffin, that I suddenly realised the dreadfulness of what had happened. We flew to Copenhagen together; I flew upstairs, Ruth downstairs. We changed planes together at Copenhagen. I went from the aircraft straight down to the baggage compartment. I told the Immigration, Customs and Airport Security people that I didn't want to be awkward but I was staying with this particular consignment all the way until it reached the baggage hold of the plane to Aalborg. On arrival at Aalborg the hearse met the plane. I rang Inge Merete to say that we had arrived and I would meet her in the morning. I went to the morgue and then returned to the hotel.

I do not remember exactly who came to the funeral, only that British Airways were marvellous. They sent everything but an honour guard. They sent a Captain, Roy Laver and his Danish wife, Else who had lived in the area; as a result, the captain spoke quite a lot of Danish. They sent the Regional Director for Scandinavia, they sent the Head of Cabin Crew Services and as I recall, a senior person from every department. I believe there were no less than five or six British Airways representatives there.

The following morning I met Ruth's parents at the morgue and the funeral procession set off for Voersaa. And then another poignant moment; as we entered the village there was a flag staff in every garden with the Danish flag at half mast. As we crawled through the village it seemed as though the world had come to a stop. Ruth was home and everybody had turned out to greet her. Through the village and on to the little church, which seemed to be in competition with the Chelsea Flower Show. The service, the burial, the awkwardness of the drinks at home afterwards and the farewell ... I subsequently collected all Ruth's things in England and took them back to Denmark and I have paid several visits to her parents in the intervening years.

I have tried to make the foregoing as complete as possible. Obviously you may want some bits and pieces amplified and there will be a lot (I hope) that you will cut out. But I thought I had better let you have everything while I can remember it. A lot of it is vividly etched in my mind. There may have been vignettes which I have missed, but nothing is embellished. I am quite sure that a great deal more will appear from the correspondence file which I am on the point of leaving home to collect, and with a bit of luck I will be able to send it with this letter.

Richard

Chapter Twenty-six

It did not greatly matter that there were 40 and not 50 expert witnesses as Ihlau
had so scornfully predicted, for none of those who were called upon to speak
following the weekend adjournment could offer anything to divert the course of
the trial.

To Weston indeed, and surely to the outcast Tasic, these men were merely
incidental agents speeding the inevitable conclusion heralded on the first day
by that headline in the local Yugoslav newspapers. 'The accused', it had
asserted, 'were guilty of [causing] the mid-air collision' – and premature or not,
this premise had not yet been substantially challenged.

Thus, Tasic could expect neither harm nor help from Captain Kroese whose
evidence was limited to that factual account of what he had seen and heard, and
how he had made his report to the persistently uncomprehending Zagreb
Control Centre.[1]

True, Tomas Zamardzic, President of the Commission of Inquiry and
Deputy Director of the Yugoslav Federal Civil Aviation Administration offered
the information that Yugoslav regulations allowed controllers to speak to pilots
in the Croatian tongue as well as in English in an emergency: this statement
barely blunted the edge of that particular accusation against Tasic. And yes,
said Zamardzic, and again it appeared almost an irrelevance, it was true that
Gradimir Tasic had been scheduled to work 50 hours in the week of the crash
and yes, that was eight hours more than the number prescribed by the
regulations.[2] Perhaps, too, there was a slight shortage of controllers at Zagreb
and it may have been the case that the radar at the airport was not properly
adjusted . . .[3]

Zamardzic's colleague on the Commission of Inquiry, Sava Zivkovic had
already made his own unfortunate impression.[4] He was, however, to cap his
avowed ignorance of certain critical aspects of his responsibility with the
equally unfortunate remark that 'the crash would not have occurred had all the
rules been respected'[5] – for all the world as if the case concerned some unique
aberration in an otherwise impeccable organisation instead of a disaster invited
by the general tolerance of a shabby inadequacy.

[1]See page 55-6.
[2]Reported in *The Times* 19 April 1977.
[3]*Ibid.*
[4]See page 124.
[5]*The Times* 19 April 1977.

It was perhaps more palatable, if no more comforting, to hear some of the other witnesses: Mladen Stojkovic, for example, Yugoslavia's chief inspector of ground flight control and another member of the commission which had concluded that Tasic had been 'overburdened'. 'Well, not necessarily', said Stojkovic ... 'not by the number of planes Tasic was handling, even though there *had* been 11 of them. If Tasic had been overburdened it must have been due to the other duties he had performed – those duties, of course, which should properly have been carried out by the assistant controller.' And which, noted the Court, Tasic had voluntarily undertaken and with no visible sign of haste or distress ... had he not even found time for humour?

Humour?
It was that small jest in which Tasic had unwittingly played the part of the straight man.
How do you spell Constantinople? LTBA ...
Thus Tasic, who had disposed of it so briefly. And it had returned now to damage him as if he, and not his colleague at Belgrade, had been the joker.

For each of the accused men the courtroom was a battlefield. For weeks and months now they had fought to remember the events by which they were threatened: and not merely to remember, but to reconcile what was known – or believed – in mind and heart to have actually happened with what could be said without fear and what, indeed, it was necessary to say in order to survive. Yet, although the strain was shared by all of them it was Tasic on whom the pressure lay most heavily, who broke, who became obviously unwell and who thus precipitated an adjournment for the next seven days.[6]

There was, perhaps, one more day of blessed respite for Tasic when the Court reconvened on 4 May since that day was wholly devoted to the examination of the hapless Lieutenant-Colonel Zivkovic and his military colleagues. It was the day, too, on which the security blanket over the interrogation of these officers so conspicuously failed to hide the story of the part allegedly played in the disaster by President Tito's aircraft.

There could be no cause for disquiet on the part of the prosecution, however, in the evidence given by the next witness, DC9 captain and qualified engineer, Zlatko Kurjakovic, who merely amplified the report which the Court had ordered him to provide.

This testimony was very largely an academic recital, confirming as it did the correct functioning of both aircraft and the acceptable conduct of their crews. But for the benefit of the Court, Kurjakovic also detailed the pilots' responsibility for maintaining contact with air traffic control during the ascent from the lower to the upper sector. He could not, said this witness, find any significant departure from this routine on the tapes given to him.

[6]There was time that day for the Public Prosecutor to change the basis of the indictment against Delic, Munjas, Dajcic, Erjavec and Pelin. See page 173.

There was, however, much more drama and a more insistent reminder of what had occurred over Zagreb in Kurjakovic's description of the manoeuvres necessary to avoid a collision, for while much of the technology of aircraft operation must have been lost on most of the Court, one extract from this evidence could not possibly be misunderstood.

> The time in which the pilot can avoid a collision is dependent on the clarity of the message from the controller and any expression he may put into it; the clarity with which the pilot receives the controller's message and the reaction time necessary for the pilot to understand and act upon that message. It depends, also, on the time it takes for the pilot's actions to physically change the position of the aircraft in space . . .
>
> . . . it depends on the action of the crew, (regardless) of any warning from the controller who has foreseen that two aeroplanes are going to be in the same place at the same time.

And in that case, stated Kurjakovic, the most effective order which the controller should give to the aircraft would be (a) to hold a certain flight level – say 325 or 327 – or (b) descend to flight level 325 or lower. It would not, he claimed, be any more efficient to attempt to change the *direction* of the aircraft.

'But how long, in fact, would it take to change the position of the aircraft *vertically*, sufficient to avoid a collision?' We can estimate that, said Kurjakovic. Allowing for the excitement of the crew within this situation:

(a) under normal conditions it should not take longer than five to eight seconds for a controller who is fit and well rested to see the danger and give the appropriate order,
(b) the time taken to react to this by an average pilot who is physically fit and well rested would normally be one to three seconds

and (c) the time needed for the aircraft to achieve the necessary vertical change of position would be seven to nine seconds.

The total time which therefore might elapse between the point of recognising the danger and repositioning the aircraft to avoid a collision is 13 to 20 seconds.

'But if, instead, the aircraft attempted to change its heading?' Kurjakovic was quite clear on this point. 'For a pilot to change his heading by 15 degrees a period of 15 seconds is required from the moment of recognising the danger: for a controller, this would take 25 seconds.'

And how much time for avoiding action had there been from the moment the pilot of JP550 gave Tasic his current height?

10.1417″	JP550	327
10.1422″	Zagreb	. . . e . . . hold yourself at that height and report passing Zagreb –
10.1427″	JP550	What height?

| 10.1429″ | Zagreb | . . . the height you are now climbing through because . . . e . . . you have a plane in front of you at . . . 335 from left to right – |
| 10.1438″ | JP550 | OK, we'll remain precisely at 330 – |

The trial had by now become an endurance test and not only for the controllers. There was no longer theatre to behold and after 12 days minds were dulled by the repetition of events, swamped by detail heaped on detail, by question upon question, by charge and implication, and above all, by the physical and intellectual effort required to find some pattern in all of this. It was difficult even to remain awake (and it was reported that the elderly juror did not trouble himself to do so), let alone alert, during the endless procession of experts who came and went, whose voices rarely rose above the level of a monotone and whose testimony appeared merely to cover ground already trampled into a confused pudding.

Thus it was Kuscer, pilot and air traffic control expert who, on 6 May, offered his opinion as to Tasic's responsibility and immediately drew an objection from Tasic's lawyer – an isolated challenge only momentarily disturbing the leaden ordeal, and incidentally initiating a fresh deluge of bewilderment for the courtroom at large.

This deluge would wash over them for at least the next three days and would leave in its wake strewn among the mass of incident, a jetsam of witnesses' names. Later, these would reappear in the Judge's findings, attached like labels to each of Tasic's separate damnations: that he was not overburdened, that he accepted the flight slip as Erjavec claimed, before JP550 called, that he accepted the co-ordination, that he gave permission for JP550 to climb without any conditions, that his actions were not sufficiently precise, i.e. in accord with the regulations, to avert the collision, that he failed to use standard English phraseology but spoke instead in Serbo-Croat ('a phenomenon of regression', said the witness Dr Marsavelski).

It seemed, by 10 May, the fourteenth day of the trial, that there was nothing left to contest, and that almost every one of Tasic's actions had been a transgression. It could not, for example, be denied that this Article or that Regulation specified the standard from which he had departed. It was impossible to deny that his had been the last human voice guiding those aircraft, or that people who had trusted him had died in so frightful a manner.

All the evidence supported these accusations, and against them, for what little it appeared to be worth, there was only – for Weston at least – the picture of the solitary Tasic on the upper sector console during that crisis which had so briefly come and gone.

And, of course, it was true that there were so many things, as the experts confirmed, which Tasic should have done differently. But then, thought

Weston, as Tatarac rose to give the final speech for the prosecution, there were so many other things at Zagreb which should have been different.

Everything, perhaps.

Everything.

It was not within the Deputy Public Prosecutor's brief that he should betray any hint of such reservations, if indeed, he felt aware of their presence. His opening sentence, therefore, was a lash laid on with the maximum vigour, accompanied by the promise of yet more pain to come. 'I ask', he began, 'that all eight accused should be found guilty and that they should be punished according to the law.'

'The trial before this Court has shown that all the statements in the Bill of Indictment are correct and that each of the defendants is guilty of causing the collision between the British Airways Trident 3 and the Yugoslav DC9 aircraft owned by Inex-Adria, resulting in the deaths of 176 passengers and all of the crew members.'

'It is a fact', went on Tatarac, 'that this accident was not caused by any fault in technical equipment or by any lack of appropriate regulations. It was caused [among other reasons which he would specify] by the hybrid practice of aircraft separation adopted by the Zagreb Air Traffic Control Centre and by the faults of the controllers which are evident from the tape recordings and which point to their practice of ignoring the rules.'

The words were given time to sink in before the Prosecutor continued: 'Nor was this catastrophe caused by the temporary indisposition of any of the air traffic controllers. Rather, it was caused by the sum of many factors: the ignorance of working rules, the overloading of air traffic controllers, the ignorance of rules governing the use of flight slips and the use of unlawful – that is, unauthorised – methods of co-operation and co-ordination.

'For all these reasons', said Tartarac, 'the Public Prosecutor holds that all the charges against the second accused, Delic, the third accused, Munjas, and the fourth accused, Dajcic, are proved. They neither acted in accordance with the rules nor did they fulfil their duties.'

In the only affirmation of dignity left to them, the three men gazed impassively ahead.

As for the fifth accused, Hochberger . . .

'As for the fifth accused, it is proved that he left his place of work without permission, that the sixth accused, Tepes did not report for duty at the prescribed time and did not properly familiarise himself with the situation in his sector.'

Four more pairs of eyes focussed on emptiness as Tatarac completed his impeachment –

'. . . the seventh accused, Erjavec, and the eighth accused, Pelin, had made many mistakes. And among them, they had not carried out the necessary co-ordination properly.'

Tatarac was about to end. He said: 'The Public Prosecutor repeats that the terms of the Bill of Indictment against all the accused are correct and that they should be found guilty and punished in accordance with the law.'

Weston looked on as his fellow representatives of the damaged parties made their own submissions in support of the Public Prosecutor, and as the lawyer for Inex-Adria requested the right to obtain damages in a civil action.

It was what had been expected and no more than a conventional and proper sequence in a public trial. And of course, his own right to make a submission had been courteously pointed out and upheld: so if he proposed to do that, and, like the others, to add his own concurrence to the prosecution's case – as the quizzical glances from the Judge and the Prosecutor now invited him to do – then it was time.

Weston motioned to Drasko, sitting beside him. Dryly he observed the other man's inner and plainly evident struggle: then Drasko rose and said heavily: 'This is not what I want to say: but it is what I have been instructed to say on behalf of Mr Richard Weston, attorney for the family of the British Airways stewardess, Ruth Weinreich Pedersen.'

'May it please this Honourable Court . . .'

Chapter Twenty-seven

To Ronald Hurst

From Richard Weston
London,

August 28th

Dear Ronnie,

I think I have dealt adequately and without trespassing on the family's privacy, with Ruth's home and the village and her parents. They were always warm and kind, and, of course, as I got to know them better we had a very sincere relationship. I suppose I visited there in all about six or seven times and I still write to them at about this time of the year. You also asked me about my own background. Very simply, I have been a solicitor since 1959 but I have always been interested in aeroplanes and had made aviation my speciality. I hold pilots' licences both in the U.K. and in the U.S. (with an instrument rating) and I fly my own plane.

We have covered an awful lot of things up to this point but as you see, this is to amplify some of the matters raised in your letter of the 19th August and especially the two points which you mentioned to me yesterday – when did I begin to act for the International Federation of Air Traffic Controllers' Associations and what was the reaction to my submission on Tasic?

In fact, at that stage – that is, during the trial – I had never had any communication with IFATCA at all. I often say now that I defended the air traffic controllers in Zagreb but, of course, I had no brief in this respect; I had never even communicated with Tasic or his lawyer, or any of them.

It was not until I got back to London after the trial that I was contacted by IFATCA representatives, and on the one hand I was thanked and on the other hand I was asked for copies of my speech and then, of course, I went to their executive board meeting in Scotland one day and we discussed the question of petitioning Tito and making an Appeal, and I said I would help in any way I could. But I was still never actually instructed on behalf of Tasic.

At the trial, the Public Prosecutor got up to make his closing address and frankly, it was a bit of a bore. It was a bore to me because I do not understand Croatian; it was a bore to everyone else because it was pretty certain what he was going to say; everybody had sat through this for some weeks and it was not as though he had great panache or a Perry Mason court presence. When he had finished, Drasko rose to speak and he started to read my speech. There was a certain amount of interest shown at the beginning, simply because he announced that 'this is not what I want to say, but I am going to say what I am instructed to say by the person whom I represent' and I was sitting there, and I had been a focus of attention during the proceedings because I used to have my bowler hat and umbrella (literally) and so it was something out of the ordinary. It did not take long, of course, for everybody to realise that the speech was not going to be a simple, straightforward condemnation of the defendants and an endorsement of everything they had been listening to from the Public Prosecutor. After a short while, more and more people turned to look at Drasko and to listen and watch intently as my theory and thesis were expounded.

There was a deafening silence as the entire courtroom, judge, tribunal, defendants and other lawyers for the defence and the families started to grasp the enormity of what was happening. I say enormity because although you and I have started to take it for granted what the Court saw then was a prosecutor making an impassioned speech for the defence and that was certainly a novel thing.

I have explained to you how it came about that the speech was made. But after the speech, the courtroom erupted into applause. The defendants, at least some of them, came over to me and started to congratulate me – I felt rather like a footballer who had just scored a goal – and then of course there was a considerable delay before the verdict, some two or three weeks later; I do not remember the date offhand. At that stage they invited me out to celebrate with them. I went to their dinner; they all spoke English, of course, as air traffic controllers and I did tell them then, very sternly, that okay, while it was possible that the speech had played some part in getting them off the hook, Tasic was still bearing the brunt of the blame and that they did not really have a cause for celebration because 176 people had died through inefficiency and incompetence, and while I could understand their jubilation it should be tempered with the thought that Tasic was not among them to share it.

Some few days after Drasko had read the speech to the Court, I said to him one day in our hotel room: 'telephone the judge and ask him whether he has got any problems on my submission or questions which he would like answered or amplified'. This was not an unusual thing to do since the judge had in fact asked for a written copy of the speech and we had delivered him four or five copies in Croatian so that he could distribute them to the tribunal. Drasko said to him on the telephone: 'are there any questions you have on the speech?' and his answer was: 'No, thank you very much. The tribunal have noted the Anglo Saxon view of the law in this matter'. That was basically the end of the conversation. But when the judge returned to give the verdict in due course he was trembling. His hand was shaking – he couldn't help knowing that the Press had been describing this as a scapegoat trial and he was decent enough to let it worry him.

And it showed. I felt sorry for the judge: but as I've already said it had all probably been quite straightforward for him until the moment Drasko began to read my speech.

Richard

Chapter Twenty-eight

May it please this Honourable Court . . .

My name is Richard Weston I have the right to make this address, specifically as the attorney of the mother, father, and two sisters of one of the crew members of the British Airways Trident Zulu Tango which was lost together with the lives of all on board in the mid-air collision near Zagreb on 10 September 1976, which said collision is the subject matter of this trial . . .

Although my official capacity here is as stated in the role of an aggrieved or damaged party I am also concerned with other matters and when I have spoken of them, I hope that the Court will find it possible to widen its consideration of this case.

What are these other matters?

Firstly, I am the attorney/administrator of the estate of the deceased crew member and therefore represent her posthumously.

Secondly, I myself am a regular passenger on scheduled airlines and as such I may be said to represent any other members of the fare-paying public who may wish to be associated with my remarks: and thirdly, I regularly pilot a private aircraft myself and am therefore dependent in that capacity for my own safety on air traffic controllers and their systems.

Fourthly, and finally, there is my role as a lawyer. While my first duty is to the Court and my second to my client, I share the overriding duty which the legal system itself owes to society as a whole. With respect, therefore, I request that my forthcoming submissions should be treated as having been prepared in order best to discharge that duty; and that, in making these submissions, I should be regarded, not so much as a co-prosecutor for a damaged party, but rather as *amicus curiae*: a friend of this Court, who desires his words to be interpreted in that light.

There are certain facts about this case which appear to be common ground between prosecution and defence and on which I do not intend to dwell longer than is necessary.

Thus, it has been established that on the 10th day of September 1976, at approximately 10 hours and 14 minutes Greenwich Mean Time (11.14 local time) a collision occurred in clear skies between the British Airways, Trident Three, Golf Alpha Whisky Zulu Tango and the Inex-Adria DC9 Yankee Uniform Alpha Juliet Romeo about 30 kilometres north-east of Zagreb VOR near the town of Vrbovec at approximately flight level 330 or 33 000 feet, ten thousand and fifty metres above sea level ...

We know, too, that the said collision resulted in the total destruction of both aircraft and the death of 176 passengers and crew members on board, that is, 54 passengers and nine crew on the Trident and 108 passengers and five crew on the DC9: and we know that both aircraft, the DC9 *en route* from Split in Yugoslavia to Cologne in Germany and the Trident *en route* from London, England to Istanbul in Turkey, were immediately before and at the moment of impact with each other (a) operating normally and under the control of their respective properly qualified crew members, and (b) operating in accordance with the directions of the Regional Flight Control Centre, Zagreb.

What else is known?

Surely, that until shortly before the collision, the Regional Flight Control Centre Zagreb had itself operated normally, controlling daily and safely, compared to any similar centre in Europe, one of the busiest air corridor networks in the crowded skies of Europe; and that the centre was, at all relevant times, up to and including the time of the collision, directed, controlled, staffed, manned and operated by, among others, the eight defendants named in the indictment.

It has not been alleged that any equipment, or crew malfunction, or failure in either aircraft contributed to the disaster, nor that any piece of ground equipment was defective, though full use of radar facilities may not have been made owing to their need for final adjustment.

Then, if all this is agreed, what remains?

... that there was confusion, misunderstanding, mishandling, breach of procedural rules and departmental regulations, overwork, unauthorised absenteeism and shortage of skilled controllers in the control centre immediately before and at the time of the accident, would seem to be undisputed, though the questions of degree and responsibility are major areas of conflict in this case. What is not disputed is that it was these factors which led directly to the accident: and, if that is so, then this Court is concerned, not with cause, but with blame.

There was no change in Modrusan's voice. Shut out by the language, Weston could only worry. Inwardly he fretted: I hope he's putting this over.

Now, I do not intend to usurp the functions of any of my learned friends in addressing myself to the evidence. The Court has already directed its own enquiry on these matters, and the issues I wish to raise are not in conflict with that function.

But, with respect, I would ask you to consider not only your power over the defendants but also, what I believe to be your responsibility to society in general and the international aviation community in particular.

If, as a result of the findings of this Court, rulings and recommendations can be made, directions given and penalties imposed which will combine to minimise or indeed avoid, not only in Zagreb, not only in Yugoslavia but in the aviation world as a whole, the possibility of a recurrence of this tragic event, then I respectfully submit that this Court will have erected a fitting monument to all those who on that day lost their lives, in the form of a safer system for those who will fly in future.

I know, and respect the fact, that this is a criminal trial. Yet I hope to illustrate that you may better serve society if, instead of limiting yourself to the punishment of the defendants, you use your powers in an imaginative and resourceful way, namely, to bring about the changes in the aviation system which this tragedy has shown to be necessary. For these men are themselves the victims of that system: not a local Zagreb system, not even a national Yugoslav system, but an international system which is undermined by its own complexity and shortcomings.

The 10th September was a black day for the aviation community as a whole. That it should have been Zagreb however, which will always be associated with it, derives partly from an interwoven chain of consecutive events and misunderstandings which on that day became public property because of the happening of one contingency in perhaps ten million; who knows the exact odds? Yet, on 10 September, instead of missing each other by but a few feet, both the Trident and the DC9 claimed occupation of precisely the same few cubic feet of airspace. That is all it was: a wing tip of the DC9 and the nose of the Trident, a few cubic feet of air for no more, quite incredibly, than a fraction of a split second.

A few feet in a million miles; an instant in time . . . and but for the coincidence there would have been an 'incident', instead of an accident: and a mere sigh of relief, perhaps, as the air-miss report was consigned to the files.

How rare, then, is this case? And what lessons does it hold for us?

Last month there were two near misses on one flight over Spain. Other errors by pilots and controllers the world over have been buried as air-miss reports instead of rearing their heads as indictments, memorial services and funerals. So this terrible possibility is not in the least a rare thing; and when the responsibility for the crash of 10 September was thrust upon the air traffic controllers, controllers the world over said 'There but for the Grace of God go I.'

An airline disaster generates worldwide publicity in a peculiar way: the public need a scapegoat, the news media consider it their duty to provide one and in an aviation disaster it is nearly always the pilot who is blamed. Instant judgement is pronounced by persons who in most cases are completely unqualified to comment and who nearly always do so before they are in possession of much more information than a rough idea of the number of people killed. It is typical of this that after the recent Tenerife catastrophe, television reporters, who were unable to interview people on the spot, actually televised an interview with the booking manager of an airline office in California on *his* view of the cause of the crash . . .

It is known, in aviation, that experts may take months to unravel and probe the mysteries surrounding an accident; but the media, thirsting for sensation, provide this information within hours, and are in no way embarrassed by the need to amend their reports almost as soon as they are printed. And indeed, they continue to amend them as second and third hand reports come to hand, or whenever some other unofficial 'official' propounds his views.

We have seen that situation here: for in this case the defendants were charged, tried and convicted by media reports on the same afternoon as the accident occurred, and when this trial started one newspaper said – and I quote – 'the accused are guilty of the mid-air collision'.

Such reporting is not only inconsistent with the defendants' rights to justice: it is also contemptuous of this Honourable Court, the effect being to predetermine in the minds of the public what the verdict should be before this Court has had an opportunity of hearing the evidence and the arguments.

It would be safe to say today that the public already expect this Court to find guilty and sentence these defendants, because they have been so conditioned by the news media: and – and this is the real problem here – that while the public can accept human failing and will be satisfied as long as the blame can be laid fairly and squarely on one or more individuals, what it cannot so readily believe is the fallibility and the failure of the system which I suggest is the chief factor in this case.

I respectfully urge this Court to recognise that, in spite of all the reports about controllers causing the disaster, the eight men who stand before you are innocent of any charge until you, the members of this Court, determine otherwise – that is your prerogative and yours alone.

It was, reflected Weston, almost as though those in the courtroom had slowly become aware of an intrusive counterpoint. They had listened, or had appeared to be listening, as the reluctant Modrusan toiled away at the unfamiliar phrases, plainly accepting that this was yet another rendering of the prosecution's theme: but among them now was the sense of a subtle change in that forceful current and the recognition of the first, faintly disturbing notes of a quite different strain.

It is interesting, Modrusan continued, that in the final analysis it is, contrary to popular belief, more productive to recognise the failure of the system and the reasons for that failure, and, as I invite you, to direct its overhaul, than to blame an individual; for if you blame an individual and lock him up, the deeper underlying problem remains unsolved and next week another individual within the same fallacious system, may make the same error with equally terrible consequences. This is the state of affairs which must not be allowed to come about and which this Court has the power to prevent.

In any civilised society where supremacy of the law is to be maintained, the community which it serves can only respect the system if its decisions are just and fair: for after all, that, in essence, is what justice is all about. And not only the methods of administering justice – which I, as a stranger to Yugoslavia, have been privileged to witness here – but also the determinations and sentences resulting from that system must be such as will be applauded, and not deplored.

To revert, therefore, to basic questions; this Court must satisfy itself that the commission or omission of an act by one or more of these accused was a criminal offence. . .

One man is accused of having been late for work. Is that a crime? Another left work early. is that to be called criminal behaviour? Then how many more criminals, I ask, are present in Court today? That early departure or late arrival may indirectly have contributed to a disaster quite out of proportion to anything which those defendants could reasonably have foreseen as a consequence of their action. Justice itself may require departmental disciplinary proceedings – a loss of privileges, even perhaps demotion and a reduction in salary: but hardly imprisonment . . .

There was no longer room for doubting the meaning of what was being said. That minor strain was becoming more insistent, impinging not only on the courtroom at large, but also now, on the increasingly incredulous controllers. It was by no means what they or anyone else had been expecting: but this Englishman, this foreign lawyer, who sat with that other army arrayed against them, seemed to have special reasons of his own for saying something in their favour. It could offer little hope of relief, perhaps: but they could at least raise their heads for a space: *something* was being said in their favour . . .

I wish, said Modrusan, to touch on another matter. The Court has heard discussion as to whether the pilots could have seen the impending disaster and perhaps even avoided it . . . one can discuss this at great length, but it might be unnecessary and irrelevant to do so. Suffice it to say that the closing speed of the two aircraft probably exceeded 1500 kilometres per hour: the muzzle velocity of an average rifle bullet is between 900 and 1000 kilometres per hour, that is about two-thirds of that closing speed. One is prompted to ask whether there is present in Court anyone who could dodge a rifle bullet? Admittedly, the matter may not be quite that simple, but at least it does start to put that particular aspect in perspective.

It is a picture, perhaps, to help you to understand the reality: and I would like now to show you a similar picture so that you may understand, too, what it means to be a controller. To do this I will take the Court into that mysterious, electronic world in which air traffic controllers live.

The Court may know that a recent study of air traffic controllers at O'Hare Airport in Chicago, by the United States Federal Aviation Administration, disclosed that over 66 per cent of those controllers suffer from peptic ulcers, a

145

condition often resulting from stress. Indeed, the team of researchers stated that the most common sight in the control room was a bottle of antacid tablets on the controllers' desks.

The job which these men do is known to be ultra-demanding and it is for this reason – as this Court has heard – that a controller is only on duty for two hours at a time; that is, he serves for one hour as assistant, and for one hour as controller: he then takes an hour's break.

Now there is no one in this room who works under such conditions. There is no one in this room who works under pressures of this kind and there is certainly no one in this room who is expected to make decisions under the same circumstances. But before you consider anything else, let me tell you something more about those circumstances.

The job of juggling airliners and making snap decisions on which hundreds of lives depend exacts a heavy price in stress-related diseases, nightmares and acute anxiety. In just the last year seven men have been carried out of the O'Hare Airport traffic control tower on stretchers, victims of acute hypertension. Most controllers there have already succumbed to one or more by-products of prolonged stress, ulcers, high blood pressure, arthritis, colitis, skin disorders, headaches, allergies and upset stomachs. It takes a special individual to withstand this onslaught day after day, because it is a job that few people could perform at all, let alone endure . . .

Of the 94 controllers and trainees at O'Hare Airport only two have been there more than ten years; most do not last five. The gruelling pace, the split second decisions – such as that which Tasic was required to make on 10 September – and the constant threat of mid-air collisions all take their toll and men leave. It is becoming increasingly difficult to entice other experienced controllers to replace them and this can only heighten the work load and stress for those who remain. The real danger here – and this fear has been voiced again and again in the aviation community since 10 September – is that the effect of a gaol sentence on Tasic, Erjavec, Pelin, Hochberger and Tepes will be disastrous for the air traffic control system generally and will directly affect the safety of the travelling public.

The room was intent now, on the speaker. By Modrusan's side, Weston noted the shifting moods as his argument developed: studied concentration from the Judges: abstracted stares . . . and on the faces of the controllers, something which had not been there since the very beginning of this trial.

Many people, said Modrusan, have difficult jobs. Many of us have stressful jobs, and many of us carry heavy responsibilities. But I ask this Court to recognise that the air traffic controllers' responsibility is a very special one and that the stress which that responsibility induces is of an abnormal nature. It is essential to view the events of 10 September in that light if these defendants are to have justice.

146

In 1973 two researchers, one from the University of Michigan and one from the University of Boston, compared the Federal Aviation Agency medical reports of 4000 controllers and 8000 pilots. Not only was hypertension four times more common among controllers: it also developed at an earlier age, and was especially prevalent in busy areas. In addition, twice as many controllers suffered from peptic ulcers.

It is also a fact that, since 1970 alone, more than 35 controllers have been permanently removed from their jobs for medical reasons. I am quoting American figures because I happen to have these available: but I am quite sure that similar situations exist in other places and other countries and the sooner they are recognised, the better for the safety of the travelling public.

A former President of the American Academy of Stress Disorders, Mr Richard Grayson, has compared the symptoms of air traffic controllers with railroad despatchers and sonar operators on nuclear submarines: and, he says, the stress symptoms of air traffic controllers are much more severe, aggravated as they are by overwork, high-density traffic and the constant fear of mid-air collisions.

A fear of mid-air collisions . . .

We are here today because of such an event, because that fear was realised: and because, it is charged, there was a failure of the human link in the system. Let us look at that system: it may be that we shall identify quite another kind of failure.

If, as happened on 10 September, the system breaks down, how much more important, I suggest, that it should be examined. It may be that we shall have to change some of the people working within it, because they cannot meet the high standards required: but to go further than that would require the most careful consideration. For some years now there has been a massive international effort underway to improve safety in the air. This effort can only succeed to the full if every airline, every agency, indeed every government and every system which forms part of the international aviation community or upon whose decisions that community rely, will combine together and operate in harmony one with the other.

What is this international interdependence?

The Court may judge what national decisions mean in the context of civil aviation: only this month, May 1977, the Spanish Government announced the installation of improved air traffic control equipment throughout the country after allegations that its airports and airspace were dangerous. An Air Ministry statement in Madrid said that the Government was 'fully conscious of the importance of the security and fluidity of air traffic and had authorised an unspecified grant to pay for the necessary improvements. These include a national network of new radar installations belonging to the Military which will be handed over to civil aviation next month'.

147

I suggest that if a similarly dramatic change were to come about in Yugoslavia as a result of the decisions of this Court, then the lives of those killed on 10 September will not have been sacrificed in vain.

For a moment Modrusan was quiet. Then he said:

In both the United Kingdom and in the United States, systems are now operating which are designed to bring to the attention of the authorities the mistakes which cause today's incidents: because recognition of these mistakes can enable action to be taken to prevent tomorrow's accidents. The American system is of particular interest because it has been in operation longer than the English system and because air traffic controllers have participated to a very large extent in that system. It involves the voluntary reporting by pilots, controllers, engineers and others concerned with safety, of mistakes, malfunctions, errors, and incidents, etc. to the National Aeronautics and Space Agency, more commonly known as NASA, who act as an intermediary between the party reporting the incident and the Federal Aviation Agency who license that individual – and upon which agency that individual depends for his livelihood. In this way, the Federal Aviation Agency is becoming aware, daily, of the problems: and they are able to take timely action when they see that a particular pattern developing in any particular field of activity gives rise to apprehension and concern.

But I ask this Court to note that the reporting of incidents is on a voluntary basis to assure those concerned that the important thing is to find out what went wrong, and to take steps to prevent it happening again. But it has been recognised in implementing the system that it would be wholly counter-productive to safety if a voluntary report were to find its reward in a prosecution. In most cases therefore, immunity from prosecution is granted – the object being to promote safety and not punish offenders.

It is not difficult to see the relevance of that procedure to this case. There is a great danger that this prosecution will undermine the international system which we have found to be necessary and which is being adopted more and more widely. It has operated in Australia for 20 years or more and, as I have said, is now operating in the United Kingdom.

It is my earnest wish that the judgement of this Court should not place that system in jeopardy, and if it is not already plain how this might happen, allow me to draw a parallel: if we were to say that doctors prescribing the wrong drugs will be subject to criminal prosecution for endangering life, what kind of medical profession would we have tomorrow?

The answer is painfully clear.

That is a parallel, as I have said: but if one looks for them, there are others . . .

Chapter Twenty-nine

. . . If one looks for them, said Modrusan, there are parallels for nearly every event of 10 September . . .

Let us take that close call, for example, near Detroit in November 1975, just after one controller at the Cleveland Air Route Traffic Control Centre relieved a colleague due for his coffee break . . . for it was the belief of the Federal Authorities investigating that incident that the relief controller may not have had sufficient time to familiarise himself with the air traffic situation before he took over. On another occasion just outside Chicago, on 5 December 1975, two controllers working side by side at the Chicago Centre Air Traffic Control Facility in Aurora, Illinois, cleared their planes into the same airspace at 7000 feet without informing each other: and just an hour later there was another Chicago near miss when a Trans World Airlines 727 jet was put on a collision course with a United Airlines 727 jet . . . in this case an Aurora controller was found to have been distracted by the fifteen other planes already in the holding pattern for the O'Hare Airport and several others approaching it. He failed to notice on his radar screen that the two 727s were closing. Fortunately the T.W.A. pilot saw the other plane in time and took the action necessary to avoid a crash . . .

Parallels . . . we are not short of parallels:

On 25 February 1960 a collision occurred between a United States Navy DC6 and a DC3 operated by Real Aerovias, a Brazilian Airline.[1] The accident claimed a total of 61 lives. At the time of the collision and for some five minutes before, both aircraft were under the direct voice control of the Rio Approach Control Centre. The same controller made all transmissions to both aircraft using the same frequency for both, but using Portuguese for communication with the Real DC3 and English – from a phrasebook – for the Navy DC6. All the evidence proved conclusively that use of two different languages to conduct Rio air traffic control was a major cause, if not the primary cause, of the accident.

Now we do not know – we cannot know – just what part was played on 10 September by the use of two different languages. But as you have heard, that aspect, like so many other facets of that disaster, was not unique: and if that was not unique, and all the other hazards were known, how have authorities dealt with these problems?

[1]Reported in Hurst, Ronald *ed.* (1976) *Pilot Error*, 1st ed., 154. London: Granada.

Mr John Leyden, President of the Professional Air Traffic Controllers Organisation of America has said that, 'historically, the authorities have only reacted after a disaster or a calamity in the air. They have not planned for it, nor have they taken the proper steps to forestall the accident from happening. We got radar coast to coast after the first controlled mid-air crash in this country, the Grand Canyon disaster of 1956. We then established and got a network of radar coast to coast. We got mandatory radar procedures after another disaster in New York, a mid-air collision in 1960 in New York. We got digitised radar read-out after another mid-air collision in Carmel, New York, where two planes had to hit before we got relief: and it seems as though the only time we get attention focussed on a problem is when there are bodies strewn on the street. There is public pressure, Congressional pressure exerted on the Agency to take action. They react: they do not plan. ....'

They do not plan. Then let this Court decide that 10 September should be the disaster that makes planning necessary, and that it will not solve any problems to imprison those persons who were guilty of human failing on that day.

We should, I submit, also be clear that this would be a most dangerous precedent. The World Airline Accident Summary published by the United Kingdom, Civil Aviation Authority, lists 76 mid-air collisions on a world-wide basis from December 1946 up to March 1975. This does not include small aeroplanes, but only refers to mid-air collisions in which one of the aircraft was a commercial airliner.

I have been unable to find, either myself or questioning colleagues, any criminal prosecution of the controllers in any of these cases although it is clear that controllers were held to be responsible in many cases. It goes without saying that in nearly every case pilots who may have been at fault were killed in any event ...

It is a fact that the aviation community, the scientists, technologists, designers, airline and airport operators and administrators have created a system which has grown into a monster before we have had time to learn how to control it. Every year there are more new airlines, more new planes; thousands of them, faster and bigger, carrying more people but all having one constant factor, namely the space in which they operate. Space has *not* grown any more accommodating and it is true that today's jets, Concorde cruising levels excepted, use the same airspace which only a fraction of them used twenty years ago.

This Court will appreciate that the ingredients of tragedy are all there. They are there in the congested air lanes, in the immense speeds of the aircraft, in the critical nature of the separation required and in the fallibility of human beings, with their potential for error and disaster. And when, as on 10 September, disaster strikes, is it really fair we should take those persons who happened to be at the centre of the system at a moment in time and visit our revenge on them?

Let me make myself clear beyond doubt. I do not for one moment ask this

Court to condone or overlook findings of culpability among the defendants, but what I do ask is that such findings should be kept in perspective. I suggest that we must as a responsible society differentiate between simple human error, culpable negligence, recklessness and criminal activity, and that while I believe in the imposition of a punishment which will fit the crime, as we in the European continent understand that phrase, we must first be sure that there has, indeed, been a crime: and in general, an act or omission said to be criminal must be supported by a criminal intent.

It is not my own intention, I assure you, to review the evidence, with a view to suggesting an apportionment of blame between the eight defendants, because there are other advocates here more properly qualified for that role than I am: but I nevertheless most strongly invite this Court to say that there was no criminal action by any of the eight defendants in this case. I respectfully submit that it would be impossible to conclude otherwise, and I venture to go further. If the law is such as to interpret their actions as criminal then I invite this Court to say that the law is wrong and should be amended. It would not be the first time and it most certainly will not be the last. I realise that that may not be an easy thing to do but where liberty of the individual is concerned the way has not been traditionally easy anywhere. If the Court accepts that submission then it would seem that maximum, or indeed, near maximum penalties for the indictment, as drawn, become inappropriate: . . . and if it should be accepted, then the Court might also consider whether the defendants' culpability for the collision was in a particular instance reckless in the sense of extreme carelessness, or negligent to a degree that amounted to a betrayal of duty; serious enough to be sure, but still short of criminal activity.

It was quite impossible, thought Weston, to gauge how much of this was going home. Certainly the Court appeared to be concentrating on what was being said and certainly the effect of this unlooked for intervention was readily visible in the sustained alertness of the accused men and the busy pens of the Press. But for the rest, from the Judge and his associates, to the Counsel and beyond them, to the seated members of the public, there was only an unrevealing blankness which could have meant anything. It might indeed, have meant that they really understood what was involved, what kind of lethal juggling was represented by the controllers' trade. Or it might, equally, have meant that the whole trial was so steeped in boredom and mystery for them that there was really nothing to do but sit through it until these fellows, who had so obviously failed to do their jobs properly, got whatever they deserved, so that they could all go home.

It was quite impossible to know: but perhaps something would get through. For all his misgivings, Drasko was doing his part: that was one thing, anyhow. Weston repressed a smile. It sounded as if Modrusan had really got into his stride . . .

*

... It is recognised by the aviation community that there are inadequacies in the air traffic control system itself. They are in fact many, most of them the result of technological deficiencies and economic considerations.

The Court has heard evidence of 700 000 flights under the control of the Zagreb Traffic Control Centre, in the space of five years, with a total of 166 complaints – that is an average of one complaint approximately every 4216 flights. I am tempted to ask how many people in this Court today can boast of being right 4216 times for every single mistake?

We are dealing here with a case based on charges of human error. In view of that it might be agreed that that record is not a bad one by any means, but perhaps it might be better understood.

Modrusan looked up from his notes. 'Let us put Mr Tasic in this situation', he said.

The people operating the system are required to make many decisions, as was the defendant Tasic on that day, affecting aircraft safety and expedition. The high speed of aircraft operating in all planes, speed, rate of climb and descent, change of direction, demands that such decisions must be made quickly and accurately. Unfortunately, as we have seen, it is this speed together with the number and complexity of aircraft in a small sector of airspace which may preclude a wrong decision from being corrected in time to avoid a catastrophe. Yet decisions made under the pressure of fast moving and momentous events may not always be the best or most efficient ones. Delay or even lack of decision can serve, as on that day, to exacerbate a complicated situation, causing a lowering of overall efficiency and consequently the level of safety. Decisions in these circumstances can only be safe provided they require the minimum of consideration in thinking time and calculation. They should follow simple and constantly practised patterns and should be accepted variations of a known pre-planned pattern of operations within a relatively simple, standardised system – simplicity and standardisation are recognised necessities and constantly recurring themes in all aspects of air operations . . .[2] yet for Tasic, that situation was neither simple, nor did it conform to any pattern he had met before.

There are many aspects of current air traffic control which fail to satisfy these requirements and each represents special difficulties for the controller. The need for a common language for aviation is readily apparent: but even in speaking our native languages there are hazards. It is a sad fact of life for instance, that incoming messages do not reach us at convenient intervals but arrive irregularly and often just at the wrong time.[3] This fact is critical in this case because research has shown that man possesses only a single decision-making channel and that all information must be passed sequentially through this channel – for example if two items of information arrive at the brain at the

[2]*Ibid.* Martin, Philip. 'Air traffic control factors', 152.
[3]*Ibid.* Allnutt, Dr Martin. Chapter 2, 68.

same time one must wait until the other is processed.[4]

There is a classic laboratory test in which different messages are put into the right and left ears at the same time – the test has shown that when the listener attends to the input to one ear he can tell his questioner virtually nothing about the message arriving at the other ear.[5] This would be the same whether the message came to the ear, in Tasic's case, from the DC9 on the radio, to his eye from the images on the radar screen, or from the slip which was handed to him. The fact is, and it is important that this Court take judicial knowledge of it, that man can attend to one thing at a time only. Enquiry after enquiry has found that the cause of an accident has been the crew's preoccupation with something else. The limited capacity of man's single decision channel means necessarily that when situations arise in which even though all the component parts of the system are working well, there is still so much information that the channel becomes overloaded.[6]

The pace and stress of the environment has made many people, not only in air traffic control, all too familiar with situations in which they are required to attend to too much input. Familiar too are the number of well-recognised techniques which human beings adopt when this overload situation occurs. Depending on temperament and capacity individuals may deal with each piece of information quickly and badly or may concentrate entirely on one source of information to the exclusion of all others. They may confuse information obtained from two or more sources or may even seek to escape from the situation by ignoring all the input, possibly by indulging in a totally irrelevant activity.[7]

There is another phenomenon, continued Modrusan, which makes an important contribution to human error and in this context it deserves special attention. The radar controller expects to see little blips moving about on the screen: he does not expect them to meet or, if they appear to, he knows that they are at different levels anyway. He sees blips, and to him, that is what they are; he could not cope with the psychological burden of regarding them as so many hundred human beings in his immediate care. He has had long experience of events such as blips passing on a screen and the fact that this has always happened in a particular manner generates a strong expectation in him that it will continue to happen in that manner in the future ...

Now, if we remember that, if we add to that the human need to make sense of situations, it is not surprising that when we are passed a slip of information saying that another plane is climbing, uncleared though it may be, into our sector, we do not neglect it, we do not cry for help; we are conditioned not to

[4]*Ibid.* page 68.
[5]*Ibid.*
[6]*Ibid.* page 69.
[7]*Ibid.* Chapter 6, 69–70.

cause panic or to become excited: instead, we try to structure it into a habitual situation . . .

I have tried, said Modrusan, glancing briefly at Weston . . . to show the Court something of the physical and psychological difficulties under which the controller works. There are, of course, many more problems than I have been able to describe here. But you have heard now, how tenuous is the human control of this immense activity and I am sure now, too, that you will understand how vulnerable are its servants.

On 10 September the final link in that system failed. That final link was Gradimir Tasic and we should keep in mind the manner of his failure. In just the space of seconds he was meant to process a situation in which two aircraft, not seen to the naked eye, but only as dots on a screen, in a semi-darkened room, were moving towards each other at 1500 k.p.h.; he was expected to talk to the DC9 pilot, keeping watch, of course, on nine other planes all fighting for a position in the queue for his single decision channel; determine the height and direction of the Trident, assess their closing levels, positions, speeds and times, and finally, to talk to the DC9 pilot in a foreign language about the whole situation – all I repeat, in the space of a few seconds. It is clear from the tape recordings presented to this Court that he tried: but in the event he proved to be only human.

Who, then, is the culprit? Tasic? Or the system which demanded more than he could give?

I ask again, how can he be guilty of anything approaching even recklessness let alone crime? At best it was human error – at its very worst, there may have been a degree of negligence. But crime? I assure this Court that that is something else again.

Weston's ear registered the new note in the speaker's voice. It might not have been conviction: but it was beginning to sound remarkably like it.

. . . Can it be crime, for instance, to speak one's own language? It is one of the charges against Tasic, is it not? Yet was it not perfectly natural for anybody in his situation, in a moment of stress to revert to his vernacular? Any one of us might have done that, regulations or no regulations . . .

I want to deal further with this question of the use of the English language . . . It is a curious thing, I think, that although we can, with the passage of time and with practice get into the habit of speaking another language it is an accepted fact in human behaviour that however practised and fluent we may be in a second language, whenever we come to deal with numbers, to count or to add up a bill we revert to the vernacular. A man may live in a country other than that of his upbringing and speak its language for years: but take him into a restaurant or a supermarket, put a bill in front of him with a list of figures and watch his lips. Invariably he will count in his vernacular. Tasic is 28: for him, English is not strictly a second language but a set of sounds, phrases, expressions and coded directives which he uses in his job . . . how perfectly natural, therefore, that when he received the slip his decision-making channel cross-referenced the DC9 in a climb to the level of the Trident in the cruise. What then can we deduce was on his mind? Numbers. Numbers. Flight levels 330, 327 and an aircraft climbing to 350. His use of his native tongue could have been predicted at that precise moment. Pressure, stress and unusual situations produce reactions not usually in accordance with those expected in normal situations and that is exactly what happened to this man.

Let this Court consider these things. We are not talking of crime, or culpable negligence or recklessness on the part of Tasic in particular. Not at all: we are talking of simple, predictable human behaviour and it is notable, I submit, that Tasic's defence is corroborated to a large extent by the indictment against Erjavec and Pelin wherein they are charged with asking Tasic to accept the DC9, despite the fact – and I quote from the indictment – that 'both could see he was alone and overworked'.

Is that enough cause for concern? Because the indictment also quotes examples of negligence and breaches of discipline among the flight controllers in the past – and Tasic himself, on several occasions was shown not to be up to situations of high work load. Is it surprising, therefore, that at some time or another a fatal error would be made, as indeed it was, on 10 September?

It is not disputed that Tasic did the wrong thing, that whatever he tried to do, involved him in this tragedy. Yet that, I insist, is no crime, nor even negligence: indeed, in England, it was decided as long ago as 1879 that 'a person who takes a reasonable decision as to a course of action in an emergency or dilemma, will not be treated as having acted negligently if that decision is shown to have been wrong'.[8]

So much for Tasic. As to the others, as to the second and third defendants, Hochberger and Tepes, very little can be said that does not appear already from the indictment. The charge against them is that their contribution to this disaster was an indirect one and here too, I believe that while their actions may have shown irresponsibility and unsuitability for the positions which they held, early departure from, and late arrival for work did not amount to criminal activity.

Is this not again a tragic coincidence?

If there had been an early arrival and a late departure on 10 September by those two defendants perhaps we would not be here today and for the rest of their lives those two men, like the others, will have to live with that thought. It is, indeed, a terrible thought: but to imprison *them* would not help anyone, or do anything towards the improvement of air traffic safety. That could only be unjust, for reasons which I have already given ...

As to Erjavec and Pelin ...

Whatever this Court finally determines to have been the sequence of events

[8]Case of the *Bywell Castle* (1879) 4 PD 219.

on the morning of 10 September, it is clear that these two men were closely associated with the decision to allow the DC9 to climb into the Trident's path.

I have already spoken of the pressures under which air traffic controllers work and offered some observations on the question of human fallibility: those comments apply equally to these men.

To anyone who has listened to the evidence in this case it must be clear that whatever else is disputed, all was not well in the Zagreb Air Traffic Control Centre. It is clear, also, that part of the blame must rest with these eight defendants, with Tasic, Hochberger, Tepes, Erjavec and Pelin because of their direct or indirect dealings with the aircraft during the last few minutes before the disaster: and with Delic, Munjas and Dajcic because of their responsibilities for the overall operation of the air traffic control centre. But as I have suggested, a far larger part of the responsibility lies with the system itself.

Be assured of that. There are obviously ways and means available to the aviation authorities in Yugoslavia to discipline these controllers: but imprisonment is not the answer.

Modrusan had almost finished reading.

... It may be, he said, that the law is such that this Court must indeed reach a finding of guilty in respect of any or all of the defendants ... and in that case I urge that if my submissions are accepted, you should respond by imposing only the lightest of penalties. I have said that they will remember this experience: they have suffered enough.

There is a thread of pain which runs through this whole sorry case. We cannot repair any of that pain, but there are things which we can do – which this Court can do – to honour Yugoslavia's obligations to the international civil aviation community. One of those things is to understand the anxiety felt by controllers all over the world as they await your verdict. Another is to remember that if a precedent is set here – if imprisonment is to be the wages of human failing – then the morale of air traffic controllers everywhere will be damaged as a direct result of that decision, and that damage may again one day cost us dear.

And, there is a final thing ...

This Court has the means to order an investigation which could lead to dramatic improvements, not only in Zagreb, not only in Yugoslavia, but throughout the world. Zagreb is in the limelight today as the architect of the tragedy of 10 September: but if this Court takes up the cause I have urged upon you, then tomorrow Zagreb, by virtue of an enlightened judgement, could provide a model for civil aviation all over the world.

Modrusan put the sheaf of notes aside. He stood for a moment then nodded to Weston and sat down: and as he did so a single pair of hands began a spontaneous clapping. The tribute was taken up immediately: it became a rising tide until the sound came from every side of the room.

Weston felt a sudden glow.

I'll be damned, he thought. Well, I'll be damned.

Chapter Thirty

It was time for Tasic's counsel, the advocates Tomo Fila and Savo Dejanovic, to make their final submission and, unwittingly, their joint approach was such as to immediately set Weston's teeth on edge.

Yes, they insisted, there was indeed such a thing as the human factor: but the Court should understand that this was responsible for only one per cent of the guilt. There was no doubt that the chief culprit had been the Federal Civil Aviation Administration for the manner in which it had managed its affairs . . .

Dismayed, Weston followed the terse translation offered by Modrusan. The defence lawyers neither lacked sincerity nor spared any effort: but all this vigour, he felt, was being wasted on a massively barricaded fortress gate while he himself had done his best to open a significant breach elsewhere. The notion that human factors might have any part to play in all this was something new in the experience of this Court: challenges to authority, criticism of functionaries and of State organisations – these were neither new, nor were they likely to bear anything but the most arid fruit.

It was, nevertheless, an impressive onslaught. The weapons were not those which Weston had chosen to use but they were by no means to be ignored: each blow would register, hopefully, with the Court and most certainly with the watching world outside. '. . . look at the controllers' working conditions: look at the lack of controllers, the long hours, the inadequacy of training . . . Look at Tasic's squalid housing – this shack in a deserted compound. Was that not disgraceful?' Surely the controllers, all of them, had done their best in those circumstances? Surely it was as the British lawyer had said, that there had been mistakes? Of course there had been mistakes, terrible and fatal mistakes: but who had committed a crime?

The defence pressed the point:

> Had the pilot of the DC9 committed a crime? During the critical moments before the collision he had not reported that he had entered the upper sector although he had had time to do that. He, or his co-pilot had been talking to someone else on an unknown frequency . . .
>
> Hadn't Tasic told him to hold his height? And, instead, hadn't the pilot continued to climb with increasing speed? And hadn't he failed to react to Tasic's warning that he had a conflicting plane from left to right?

If these were crimes, Fila and Dejanovic persisted, then what about the British pilots?

> Had they committed crimes? They, too, had not reported in to the upper sector although they were due to do so ... and what were they talking about in the cockpit? Wasn't it true that they were talking about the London [*sic*] vegetable market? that they were solving crosswords? and that they hadn't reported the overflight of a fix? Let the Court consider all this. Let the Court ask how many mistakes had been made and whether, at that late stage, Tasic's own mistakes really contributed to the disaster. The truth was that he had found himself facing a conflict situation. Bearing in mind that he was a qualified controller it was unthinkable that he would consciously risk so many lives ...

It had all been said: accusation, denial, justification, protest ... the Court had endured its fill of these things from the moment that Tatarac had begun to read the indictment and all the way through the gruelling course run by the procession of witnesses; through the hedging and parrying and sudden outbursts of the controllers themselves and through the prosecutor's summing-up and its endorsement by the three lawyers of the damaged parties – British Airways, Inex-Adria and passengers ...

If atmosphere and instinct were reliable guides that endorsement – couched in near identical terms save that the Inex-Adria lawyer had asked for the right to recover damages in a civil action – seemed to seal the controllers' fate. Tasic appeared to recognise this as he offered his own low-voiced plea.

Dejected, and no longer buoyed by that surprising sally by Weston, Tasic stated that he fully concurred with the defence provided by his counsel.

Almost, it was over: but in a final sparking of that besieged spirit he added:

> ...it wouldn't have happened if I'd had the slip. It wouldn't have happened if the crews had all reported on time ... I know it's possible that in the last seconds I might have done better – but I was taken by surprise: as it was, I did everything I could to avoid the collision. It's unbelievable that I am accused now: quite unbelievable ...

But unbelievable or not, it had happened and there was nothing more to say or to wish for since none of those events could now be undone.

The Court would announce its findings on Monday 16 May and for the few days in between, there was nothing, either, that anyone else could do. The lawyers, the officials, the Press and those members of the public who had attended the trial would return to their homes or hotels for the weekend. The controllers, too, albeit on a long leash, would likewise go home for whatever comfort friends and family might be able to provide. Weston himself would have the blessed diversion of travel: he proposed to fly home to Geneva and return to hear the verdict on Monday.

Only Tasic, as always since that morning of Friday 10 September, was accorded his own sad privilege: unobtrusively shepherded by his policeman he was returned to gaol.

Chapter Thirty-one

Only the controllers would search the opening sentences of Judge Branko Zmajevic's findings, as indeed they would search every subsequent word, for some hint of deliverance or condemnation. Those others in the courtroom for whom the verdict was less than vital steeled themselves for yet another repetition of those wearisome facts and times. ...

Briefly, the Judge reminded the Court of the origins, condition, course, flight levels and intended destinations of both aircraft, and of the established competence of their commanders. He repeated the significant dialogue between pilots and controllers, noting that the Trident had been identified on the upper sector radar by its Squawk code, Alpha 2312, and the flight level read-out, 330.

Judge Zmajevic retraced the DC9's climb after the take-off from Split, charting its entry into the middle sector and its arrival at flight level 260 where it levelled. 'Eighteen minutes after taking off from Split', said the Judge, '. . . at a speed of 316 knots, this aircraft was 62 kilometres south of Kostajnica. The time was 11.04 a.m. and at that moment the Trident crossed the Yugoslav-Austrian border.'

We know what happened, said the Judge: Adria 550 called for a higher level and the middle sector controller, the seventh accused Bojan Erjavec, told the crew of that plane that levels 310 and 330 were not available. Erjavec asked whether Adria 550 could climb to 350 and received the reply: 'affirmative, with pleasure'. 'It had been shown that permission to climb to that level was given at 11.07 and 40 seconds, that the crew acknowledged this permission and that Vienna was informed that Adria 550 would be at flight level 350.'

Step by step the Judge followed the record up to the moment of the collision: the DC9 crew had been late in reporting in to the upper sector and in fact, had not done so until 11 hours 14 minutes and 4 seconds. Their time had differed from that of ground control by approximately one minute.

The upper sector controller had immediately asked for their level and given the reply '327' had begun to speak in Serbo-Croat.

The Court heard this out, too, as they had heard so much else: yet it was the unemotional recital of that dialogue during the last few seconds before the crash which suddenly riveted the attention and reduced that mosaic of ciphers to an understandable arithmetic.

11.14 22"	Hold yourself at that height . . .
11.14 27"	What height?
11.14 29"	. . . the height you are climbing through, because . . .
	e . . . you have a plane in front of you
	at . . . 335 from left to right.
11.14 38"	OK, we'll remain precisely at 330 –

Among the confusion and complexity of this trial it was this fragment, this recurrent glimpse of the edge of an abyss which, for most of those present, truly illuminated what had happened.

It could not, of course, suffice, since whatever *had* happened was now to be carefully dissected: Tasic, as the first accused would hear how his own behaviour was to be described for the record.

Tasic himself has stated that soon after 11.00 Erjavec approached him concerning the climb of this aircraft but he waved his hand at him because he was busy . . .

After that Pelin approached him to co-ordinate the climb of that plane into the upper sector on its way to flight level 350 . . .

All this, the Judge found, was confirmed by Erjavec and Pelin: but the actual co-ordination was determined by that 40-second blank in the tape records between 11.07 and 11.07 and 40 seconds . . .

Tasic is therefore blamed for accepting that co-ordination: that is, for giving clearance for the DC9 to climb into the upper sector . . .

. . . and this circumstance, said the Judge, is fully and without doubt determined by this Court.

That absence of doubt on a wholly hypothetical transaction was underpinned by a masterly qualification: the co-ordination implied by the gap in the tapes and asserted by witnesses hostile to Tasic 'was not carried out in an irregular manner: Tasic is therefore not charged in that respect'.

Both Tasic and Pelin had agreed that the latter had pointed his finger at the radar screen to show Tasic the target of a plane. Nor could there be any confusion as to which plane had been pointed out since JP550 had been the only aircraft south of Kostajnica at that time.

. . . and JP550, of course, had already asked for permission to leave flight level 260 and climb higher. The aircraft was told to wait – at 11.06 and 14 seconds – and received that permission, according to the tape at 11.07 and 40 seconds . . . clearly, after the co-ordination between the middle and upper sector controllers had taken place. Tasic heard that reiteration, sensing that this plank was not only to be swept away from him: Judge Zmajevic was now about to lodge it far beyond his reach. '. . . and had not Tasic himself told Pelin that the DC9 could start climbing immediately to flight level 350 if he was already at 310? Or that if he couldn't make 310 before Zagreb he could climb after that?'

'You lie!' Tasic had said to Pelin. 'You lie!' But if he *had* emerged the victor from that confrontation, that too, was now to undo him. '. . . this shows, does it not, that Tasic had information about this plane and that the co-ordination must have been carried out? Clearly, it would have been impossible for him to give such conditional advice without data and without any co-ordination.'

There is the matter of the slip . . . Erjavec and Pelin are agreed that the slip for JP550 was handed to Tasic at 11.12 and 12 seconds, that is, when the plane had crossed flight level 310 and had been turned over to the upper sector frequency of 134.45 Mhz. . . .

Tasic, for his part, claims that he received the slip about one minute, or at the most, a minute and a half before the collision. Questioned a second time[1] Tasic said that he received the slip from Pelin one minute, at the most, before JP550 called: that is, at approximately 11.13 and 24 seconds. Again, he stated to an investigator that it might have been 10–15 seconds before the plane called: and in his written statement that it was during the thirteenth minute . . . It is, of course, quite understandable that Tasic cannot remember the exact moment at which he received the slip . . . but all these statements show without doubt that he recieved the slip before JP550 called.

What did this slip tell the first accused, Tasic, and what does it tell the Court?

Tasic could see that flight level 180 was crossed out because the plane had left that level. He could see that flight level 260 was circled and crossed, meaning that the aircraft had been kept at that level for a time and had then left it: and he could see that flight level 310 had also been crossed out, and that the only unmarked flight level was 350. All this could mean only one thing to Tasic: that the aircraft was in his sector and was climbing to flight level 350.

Tasic disputes the co-ordination, said Judge Zmajevic, but if he *had* been taken by surprise, what was his response to this?

Did he ask Pelin for an explanation?

Did he ask Pelin which plane this was?

Did he ask Pelin where the plane was at that moment?

He did none of those things because he *had* accepted the co-ordination. And if he had not, then he would have asked whoever gave him that slip for an explanation.

If Tasic himself had entertained any hope of deliverance at this stage it could hardly have survived the Judge's introductory statements: not only because so much that was circumstantial seemed to conspire against him but, in addition, because Judge Zmajevic was uncompromisingly blunt in his almost total dismissal of Tasic's defence. . . . Vienna, for example, had been informed that JP550 would be at flight level 350, and Vienna had received that message, as

---

[1]The Judge quoted the relevant document or witness in support of each assertion.

the tapes showed; and that, as far as the Judge was concerned, was conclusive – for why would Pelin have sent that message unless he had received permission to climb the aircraft? It was clearly impossible to believe Tasic in this matter since he had – doubtless in his desperation – given a different version of this episode on a number of occasions during this trial.

First, he had told the investigator that he had given unspecified clearances to climb 'two or three times' during his shift, despite the fact that only the DC9 had requested such permission. He had amended that statement 'almost immediately'. No, he did *not* remember whether he gave clearance for that climb: only that JP550 had been at flight level 222 'near Kostajnica' and had asked for permission to go to 350 ...

But he *did* remember calling the Trident: that aircraft had declared itself to be 'passing Zagreb and flying towards Nasice' which made Tasic assume that JP550 could climb behind the Trident. But in this particular, Tasic was telescoping events and not surprisingly, had begun to wade more deeply into a swamp of contradictions. Tann, in the Trident, had, in fact, called Tasic at 11.04 12″ and had given him One Four as his estimated time of arrival over Zagreb: Tasic had acknowledged, asked for a Squawk and told Tann to report passing Zagreb at 330. Towards Nasice was understood, but certainly not mentioned, while the alleged 'co-ordination' had not so far taken place: any 'assumptions' as to whether JP550 could climb behind the Trident must have been made significantly later. It was obvious, commented the Judge, that while Tasic 'did not remember' giving the clearance he was also searching for justification for giving it: really, it was typical of the discrepancies in this part of Tasic's evidence ... one only had to follow the record of his flounderings. ...

Thus, on 10 September 1976 he 'could not remember' whether *anyone* had asked for clearance to enter the upper sector: but subsequently, although he 'did not remember' what Pelin had been pointing out on the radar screen, he did remember that Pelin had pointed at *something*. Tasic had stated that 'something went wrong' with the DC9, that it might not have been under full power (presumably to support the suggestion that JP550 could otherwise have reached a safe height, i.e. 310 or 350, before Zagreb) and that he had contacted that aircraft 'somewhere near Kostajnica close to Zagreb: and because the plane had "special equipment", that is, appropriate performance, he believed that it could have climbed to flight level 350 in two minutes'.

And why should he think that? Why should he need to think about how fast that plane could climb and why should he remember those 'two minutes?' That was obvious, too, said the Judge. It was surely because he had given permission for that plane to climb, *well before* it actually contacted him for the first time at 11.14 and four seconds, only 38 seconds before the collision.

It was in the nightmare quality of Tasic's misery that each attempt to ward off what was being heaped on him merely sank him further into that swamp: for he had made *another* statement on 1 October 1976, in which he had claimed that he

164

had given permission to JP550 to climb to flight level 310 and remain there. This was quite ridiculous, pointed out Judge Zmajevic: what kind of 'permission' could Tasic give to an aircraft flying in the middle sector? It would be remembered that at that time, in any case, the middle sector controller had Olympic 197 at 310 and would most definitely not have accepted such a co-ordination, assuming that it had actually been offered. It was obvious, again, that Tasic was now 'modifying' his defence and that he had continued to do this.

Had he not contradicted his claim of 1 October in subsequent statements that he had *asked* for JP550 to be kept at 310 and that he had *not* given permission for the plane to begin climbing? Surely, his purpose then had been to show that during the disputed co-ordination he had believed JP550 to be at flight level 310 already and *not* that he had given permission for it to climb to that level . . . ?

As for what Tasic had actually put forward at the trial, hadn't he introduced an entirely new variation on his previous defence? He had told Pelin, he says, that the DC9 should be kept at flight level 310 and that the plane could be cleared for 350 after it had passed Zagreb.

But he had also added something he had never mentioned before, namely, that he had also told Pelin that JP550 could climb to 350 if he was already at 310 'and if he could climb quickly'.

Once again, it is clear that these statements are intended to show that Tasic had never given (an unconditional) permission for the climb . . . but, of course, the middle sector controller could not possibly agree to any of this: he could not 'keep the DC9 at 310' since, as has been remarked, Olympic 197 was at that height, flying in the same direction, while Olympic 182 was also coming in at 330 from the direction of Metlika, heading for Kostajnica and Sarajevo.

Suggestions that an aircraft – this aircraft, JP550 – might 'climb very quickly' were simply absurd. The idea was vague and unusual and in no way would have helped the middle sector controller to deal with his own traffic. . . .

No, Tasic's 'conditions' would not have been accepted by the middle sector and on this score it was only necessary to weigh Tasic's contradictory statements, from hearing to hearing, against the identical statements of the seventh and eighth accused, Erjavec and Pelin: that Tasic *had* accepted the co-ordination and that he *had* given permission for the DC9 to climb to flight level 350.

It was, noted the Judge, the only possible conclusion: and the scorn which he had poured on the notion of 'conditions' might have successfully buried that topic save that there were those in the Court, Weston included, who had now noted an anomaly on their own account: for if, indeed, Tasic had not made the conditions which were now described as preposterous, how was this fact to be squared with Judge Zmajevic's earlier pronouncements.

. . . and had not Tasic himself told Pelin that the DC9 could start climbing immediately to flight level 350 if he was already at 310? Or that if he couldn't make 310 before Zagreb he could climb after that? This shows, does it not, that Tasic had information about this plane and that the co-ordination must have been carried out? *Clearly it would have been impossible for him to give such conditional advice without data and without co-ordination.*

For the bewildered Tasic it was all part of the nightmare, as it was also part of that nightmare that the Judge quoted 'the experts Jaksevac and Kuscer' in support of his own view: yet nothing that had gone before could match the bizarre nature of what was to come next, for Tasic's every dereliction was now to be contrasted with those rules and regulations which, under the stress of events, he had so signally failed to keep in mind.

> The accused could have refused the co-ordination in accordance with Annex 11, Article 3.5.2 and Article 3.5.2.2 of the same Annex ... Further, he could have asked the middle sector for the Mode Code of his own sector to confirm the identity of the plane and its height ... But the main mistake was made at the moment when Tasic gave the DC9 permission to climb: that clearance was not in accordance with Annex 11, Chapter 3, Articles, 3.3.1, 3 and 3.3.3. and contained none of the separation minimi stipulated in Annex 11, Chapter 3, Articles 3.3.4 1/ and 2/ a/ and b/.

It was quite astonishing, thought Weston, how selectively these departures from the rules were being used. The administration's own forbearance towards its ramshackle structure and precarious operation had been dressed up in rationalisations such as the 'shortage of trained manpower' and the 'stringencies imposed by economic reality': but if this judgement was about to offer any kind of comparable extenuation for Tasic there was certainly no sign of it yet. There was, at this point, only the unanswerable charge that Tasic had not done what the rules said he ought to have done: it was a redoubtable cudgel which now belaboured him and it stirred enough dust quite to obscure such peripherals as the need to make life or death decisions in seconds, or trauma amounting to near-psychological collapse or any of the happenings which had brought Tasic to that pass.

It was all laid down.

> ... if Tasic wanted to apply longitudinal separation he had only to remember PKL, Chapter 4, Section 3, Article 402b which dealt with this situation thus: 'When JP550 reports overflying VOR Zagreb in the direction of Graz at 310 and BE476 crosses VOR Zagreb in the direction of Nasice ... and when both aircraft (are following) separate radials of VOR Zagreb, then the clearance will be given to JP550 to leave flight level 310 and climb to 350.

It was laid down and it was a tidy solution to the problem of separation and

Tasic – with all of 38 seconds in hand in which to conduct his dialogue with the imperilled JP550 – had not given any consideration whatsoever to PKL, Chapter 4, Section 3, Article 402b – or, for that matter, to the conditions of DOC 4444, Section 3, Article 3.1 which, of course, also laid down the correct vertical separation he should have imposed – 1000 feet for aircraft flying below 29000 feet and 2000 feet for those flying above that level . . .

He had been equally remiss in the matter of lateral separation since the correct procedure was covered by DOC 4444, Section 3, Article 7.2.1.3 and, for good measure, in PKL 410 (a) and (c). But Tasic had failed in this, too: according to *those* rules he *could* have said that 'JP550 can climb to flight level 350 immediately, but it must cross flight level 330 before Kostajnica.' He would also, of course, have had to ask the middle sector assistant whether JP550 would reach 330 before Kostajnica and this would all have been in accordance with DOC 4444, Section 3, Article 5.4 which stated that an 'aircraft could be cleared to change its flight level at a specific time, place and speed'.

But Tasic, who *could* have said this and who *should* have done that had not, in fact, said or done any of these things; and whatever he *had* done had been quite contrary to the provisions of PKL, Chapter 4, quite apart from his default in the matter of the aforesaid Annex, Chapters and Articles. Profusely studded as it was with equally mind-deadening references to DOC 4444, Chapter 10, 1.4.1.2./1.5.1.1/2/ and 3, the judgement continued its relentless catalogue of all the alternatives which had, seemingly, been open to the upper sector controller. He could have sent JP550 to flight level 320 and left himself with the option of giving further instructions as the traffic dictated. And, in any case, his clearance should have been precise, so that the middle sector controller could also have passed on a precise instruction.

A precise instruction?

Weston strove to recollect the wording on the tape transcript. Hadn't Erjavec given a precise instruction when he'd said: 'Adria 550, recleared flight level 350 . . .'? It was another point to be noted in case of a possibility Weston had not yet had time or cause to dwell on: but, for now, Tasic's professional failings were still being paraded.

. . . it was contrary to DOC 4444, Chapter 10, Article 2.5.1/2.2.5 and 2.5.2 . . .

But now, the significance of that repeated assurance that Tasic had issued his clearance 'without instructions or limits' was made clear: by so doing Tasic 'had brought the aircraft into a conflict situation' although 'it was easy to foresee such a situation'. One only had to look – and the assumption here was that Tasic had had time to look – at what both aircraft were doing and in which direction they were flying and note that JP550 had been given a clearance for flight level 350. The expert Kuscer had agreed that these basic elements were quite sufficient to warn Tasic of the danger . . .

That agreement, of course, had been arrived at in quite another ambience: it had not been the expert, Kuscer, who had found himself in Tasic's situation on

10 September and plainly, that concurrence might not necessarily have reflected the problems of a single-handed controller who had been abandoned by his colleague. No doubt that evaluation would be forthcoming. It seemed, however, that it was well down the list of priorities in this judgement: first came that long long column of Tasic's own errors.

> ... he had received the slip for JP550 at 11.12 and 12 seconds. Just by comparing that with the slip for BE476 on the console he could have realised that a critical situation was developing ... he did not do that ...
>
> ... he had become aware of the danger when JP550 called: but instead of taking the necessary measures to prevent the collision, Tasic remained in communication with the pilot and also regressed into his own language ... Serbo-Croatian. He had not acted in accordance with the rules, specifically DOC 4444, Chapter 10, Article 5.2/Information on Danger of Collision: and the rules of course, laid down *exactly* what he should have said. He *could* have said 'Turn left immediately, course 270' and the conflict situation would then have lasted only seven seconds. He could also have given JP550 full information on the closing traffic: that is, he should have determined the distance in nautical miles and given the crew the flight level and type of oncoming plane, if known. He should also have given them the relative speed: instead, he merely gave the direction of the other aircraft ... and of course, he had failed to use standard English phraseology in the forms laid down by DOC 4444, Chapter 10, Article 6.2.5/ manoeuvres d/ and i/ ...

Weston found it difficult to hide his irritation. The Judge's stress on rule breaking might well impress the public so long as they did not enquire too deeply into what other rules had been broken and by whom: but it also diminished this tragedy and attempted to mould it into something more manageable, perhaps, than the idea that controllers, like any other human beings, had their breaking point, or that they could carry just so much in their minds until something happened to upset the picture.

That was it exactly, he thought: Tasic had simply 'lost the picture' and had found himself living out that special nightmare which haunted every other controller. No one could explain quite how and why it happened: only that each controller handled the job in his own special way and thus made it impossible to predict the limit of his own capacity, not merely to hold that mental image, but also, to accommodate it to an incessant flow of new data.

The analogy he should have given to the Court, Weston realised, was that of the controller building up his picture of the traffic situation much as he would build a pyramid of playing cards: it could be piled up, storey after storey until suddenly, with the addition of one more card, the whole thing collapsed ...

That final card might be the arrival on the frequency of yet another aircraft, or something quite unexpected such as an aircraft taking an incorrect route, or responding to an instruction addressed to another aircraft: but whatever the reason, the effect would be shattering. That vast bank of short-term, transient information stored in the controller's brain would be dumped, to leave him groping among the chaos of that fluid mosaic he had previously regulated.

No wonder they called it losing the picture, mused Weston: those controllers who could cope with the trauma could begin again, building up the pyramid, card by card, and if it had happened during a simulator exercise, well, they could freeze the situation until it was back under control.

And those who couldn't handle it, he thought, were carried away: but quite often, not before their voices had revealed their plight to worried pilots: or they finished up like Tasic who had not found it possible to freeze those moments, save in his own mind, and for whom there had certainly been one card too many –

Meanwhile, there was Article 2.9.3.2. '. . . in case of emergency it is permitted to use a separation of only 500 feet'. Tasic had not availed himself of this provision. On hearing that JP550 was at flight level 325 he should not have paused to check the height but should immediately have ordered the aircraft to go down to 320. Or he could simply have ordered JP550 to remain at 325: no aircrew would have ignored these orders . . .

No, thought Weston, no, no, no. He checked his own notes of the transcript and found his suspicion verified: the pilot of JP550 *hadn't* told Tasic he was *at* 325. What he'd actually said – and it was quite a different situation for Tasic – was that he was *passing* 325: and by doing that he had unknowingly injected a new element of confusion into the controller's disorientation.

But this, Weston knew, was in any event an unrewarding road to travel since any dissent on this score was merely an acceptance of the terms in which the case was being judged: but now that 'possibility' which up to this moment he had resolutely shelved crystallised suddenly into a firm knowledge.

They were obviously going to be hard on Tasic and very probably on the others: but if his own submission to the Court was to mean anything at all he would have to see to it personally that they didn't succeed in burying the whole thing here, in the Zagreb District Court, under this welter of broken rules.

Perhaps, after all, he had better begin considering the next step . . . and yes, he decided: if necessary there would be a next step: let's see how it goes for him.

Chapter Thirty-two

... and if the crew *had* been ordered to 'go down' or to remain at flight level 325, said the Judge, the crew would have reacted quickly. It would be remembered that the expert Kurjakovic had testified that the time necessary to bring the aircraft into a safe position was a minimum of 13 seconds and a maximum of 20 seconds and if one counted back from the moment of collision at 11.14 and 42 seconds it became evident, did it not, that Tasic had had enough time – had he been precise, of course, as required by Article 2.9.3.2. – to order avoiding action. It would even have been sufficient if the avoiding action had been begun in the twentieth second for it could then have been averted by a normal flight manoeuvre. It would even have been possible to avoid the collision if the pilot had only begun to act in the *thirtieth* second although *that* would have meant an abnormal manoeuvre with the possibility of injury to crew and passengers and damage to the aircraft ...

But what had Tasic done with these critical seconds? He had wasted them, nothing less.

He had begun that fatal delay by asking the pilot of JP550 to repeat his present flight level: and on hearing that that was 327 he had uttered a vaguely worded demand that JP550 should remain 'at that level' without specifying the exact level number. Thus, he had caused the pilot to query what height he was talking about ...

And Tasic had begun to explain *that* and he had *still* omitted to give the flight level by number: he had failed to recognise the onset of the critical situation in time and so it was beyond doubt that this, and all of his other omissions and mistakes and departures from the Annex, Chapters and Articles as described in the Judgement, amounted to a negligence which had caused the collision. There were other aspects relating to Tasic's responsibility for the collision which would also be considered: but that responsibility, by virtue of his negligence, was conclusively proved.

So Tasic was to be found guilty, although the Judge had yet to use that word: it would come, of course, inevitably now, but its impact would be the more resounding for some further shredding of Tasic's defence. There was, for example, the suggestion that Tasic had been 'overburdened' – a term, pointed out Judge Zmajevic, which no authority had yet defined. Neither the International Civil Aviation Organisation, ICAO, nor the Yugoslav Federal

Aviation Administration had laid anything down to cover the physical and psychological limitations of the controllers' work and indeed, it appeared to be impossible to make any useful generalisation, since the component factors were subjective as well as objective. The subjective considerations included the personality of the controller, his psychological and physical condition, his skill and experience and his susceptibility to fatigue: while the objective factors included the number of aircraft being dealt with and the number and frequency of vertical movements: to these could be added the frequency of incoming and outgoing messages, the varying speeds of the traffic, the necessity of co-ordinating with neighbouring sectors, authorities or military bodies and, of course, the incidence of sudden or otherwise unforeseen changes enforced by emergencies, weather conditions and equipment damage, etc.

The state of being 'overburdened' then, repeated the Judge, was by no means measurable, and while the Commission under Judge Jakovac had concluded that Tasic had indeed been 'overburdened' that interpretation had been challenged by the expert witness Mladen Stojkovic.

Tasic, said this witness, was not overburdened by the number of aircraft he was handling, but by the additional duties he had undertaken, namely, those which were properly the tasks of the assistant controller . . . he was *not* then overburdened and the Commission should have given more careful thought to that term. The more so – and the implications behind this pronouncement were certainly clear to Weston – since the alleged overburdening of Tasic served as the basis for the case against the other seven accused.

In what way, asked Judge Zmajevic, had Tasic appeared to have been overburdened? How was it possible to use this term of a person who had voluntarily assumed the responsibilities of his assistant, who had unnecessarily extended his transmission by the use of non-standard phraseology and who, in fact, had introduced that wholly superfluous element of humour into his dialogue?

The Court did not dissent from the opinion of Stojkovic: it was a fact that the sector was meant to be monitored by a staff of two or even three, but that did not necessarily mean that Tasic was overburdened – Tasic himself, said the Judge, had not made that claim in his defence . . .

It was a curious failure to acknowledge the opening words of that defence, for Tasic had said and had been widely reported in the Press as saying, that he had been doing the work of three men. But it was, perhaps, no stranger than that wounding assertion that Tasic had been the humorist that day. But Tasic, observed the Judge, should not be blamed, as the prosecution had suggested that he be blamed, for maintaining 'normal radio communication' while he was aware of the danger. The prosecution's reference to Tasic's calmness and tone of voice was not necessarily to his detriment since the controller did not wish to create a state of panic: but clearly, under the pressure of that final minute he began to stammer, to speak in his mother tongue and, of course, to give imprecise instructions – this showed that Tasic had certainly not 'maintained normal communication' as the prosecution claimed . . .

It was another small victory for the defence, and about as significant in the scale of things as that other relief which the Judge had conceded, namely, that Tasic bore no blame for the manner in which that co-ordination had been carried out: but the Judge now entered yet another objection, this time, to the assertion by Tatarac that Tasic had known that the Trident was at flight level 330 at the moment of informing the DC9 pilot of his danger.

What height? the pilot had asked: and Tasic had said: 'the height you are climbing through . . . because you have a plane in front of you at 335 from left to right'.

Of course, said the Judge, everything pointed to the fact that Tasic knew that the Trident was at 330: this was again supported by the experts and also by the tape. BE476 had given notice that it would fly at 330. It flew at that level over West Germany and Austria and the collision occured at that level. Meanwhile, Tasic had asked for no change in height: he had made no enquiries about the height of this aircraft and he had written down level 330 for BE476 on the console slip . . .

And yet, said the Judge, none of this is conclusive: it is not possible to say beyond doubt that Tasic *knew* during those final seconds, that the Trident was at flight level 330.

It will be remembered, added the Judge, that in the course of that transmission to the DC9, Tasic had begun to stammer before saying '335' and that he had also paused at that point . . .

Why was that?

One possibility was not mentioned: that Tasic, so suddenly aware that both aircraft were doomed in the next seconds, could simply not bring himself to say '330'. Instead, he may have attempted to mask the terrible reality by 'adding' that extra 500 feet to the flight level of the Trident.

He had, however, offered his own reason for that hesitation and, said the Judge, the Court must accept that defence: that, at the moment he looked at the radar screen, he saw that the Trident was at 335. On this matter they had heard the Head of the Technical Service, Veljko Winterstaiger and according to this authority, such a malfunction was possible within the Julia system.

That malfunction could be momentary and would definitely resolve itself during the next revolution of the aerial: but the false reading would remain on the screen for 12 seconds, quite long enough for Tasic to realise that there had been a malfunction . . .

It is agreed that it is impossible to prove or disprove this aspect of Tasic's defence and the Court must therefore consider the point in favour of the accused. The statement that Tasic knew that BE476 was at flight level 330 will consequently not weigh in our deliberations: but this Court considers that the accused, because of his actions, was negligent . . .

. . . and Tasic is therefore found guilty.

Chapter Thirty-three

Delic and Munjas. . . .

Both of these men, said the Judge, were responsible for the efficiency of the Zagreb Flight Control Centre. Both were charged with laxity, with an irresponsible supervision of public traffic according to Article 272/1 and 3, to the extent of a conscious neglect of rules and regulations. This charge also required that the Court should determine a causal link between the disaster and the activities of the accused . . .

It had been noted that the Public Prosecutor had decided to change the indictment against these two men. Their offence was now the failure to organise the work of the sector according to the regulations, to supervise the work load of the controllers, or to require a traffic and duty schedule from the chief of shift in order to monitor the controllers' work load: and according to the accusation, Gradimir Tasic had been overburdened between 11.04 and 12 seconds and 11.16 a.m.

But the Court had determined that Tasic had not been overburdened: and that being the case, this charge had no substance. There could be no causal relationship between the collision and this aspect of Delic's and Munjas's activities. If that question were to be put aside, said the Judge, what of the suggestion that it was possible for the accused to check the controllers' work load or even carry out that task by analysing the flight schedules?

The Judge did not believe this to be feasible. Every day, he said, there were over 400 aircraft in the region representing only the regular airlines. To these could be added the special domestic charters, special foreign flights, the aircraft belonging to flying clubs and, of course, those of business aviation. And finally, there were military flights.

All this traffic was made even more difficult to analyse by the incidence of factors such as changes in flight schedules, earlier or later arrivals or departures, bad weather and diversions from airports . . . these values were not fixed and it would not be possible for whoever was in charge of a shift to come to valid conclusions. It had, in fact, been suggested that the analysis of schedules should be performed instead, by specialist groups yet to be established. This suggestion was supported by the experts Kuscer, Jaksevac and Stojkovic: Stojkovic, however, had given it as his own opinion that even such a specialist group would probably only be able to predict 50-70 per cent of the traffic correctly.

Nor was it practical for the accused to have 'limited' certain operations. This was also suggested in the indictment and it was, really, quite difficult to understand what was meant by this. It could only apply to the relief of a sector and to achieve this it would be necessary to manipulate flight levels – which, of course, would affect all other traffic – the time of entering the flight information region – which, according to ICAO, required three hours' notice – and route changes, which were also not permitted, unless by the military.

Some countries had re-routed flights, and some had attempted to establish special routes at weekends . . . but this had not been tried in Yugoslavia . . . All this, the Judge repeated, led to the conclusion that there had been no conscious neglect of rules and regulations on the part of Delic and Munjas.

On the matter of discipline, Munjas had quoted the record of disciplinary action which Delic and himself had initiated against a number of controllers, including Tepes, Tasic, Hochberger and Pelin. The punishment had ranged from compulsory re-training to demotion and dismissal and it was difficult to find a single controller who had violated the rules and who had not been punished. This charge, therefore, was also baseless . . . It had additionally been charged that Delic and Munjas had been in default because they had not sufficiently instructed their flight controllers, because they had tolerated the use of unauthorised procedures by the controllers and in respect of their lack of supervision of the transfer of control from sector to sector . . . None of this was valid. These charges had been disproved either by the demonstrated performance of Delic and Munjas or because they were not required to issue instructions in the cases cited . . .

Indeed, said the Judge, the accused had actually done more than was required of them. There had been discussion on the necessity to provide a flight strip for JP550 . . .

Firstly, a strip was not made out specifically for the sector, but for Flight Control bodies – for the Air Traffic Control Unit, the Area Control Centre, the Approach Control Office and for Airport Control. The sector was not such a body . . .

When a strip was made out by the appropriate assistants it was handed over to the controllers of the sectors specified in the flight plan. There had been no need to provide a strip for JP550 in this case, since the aircraft had not been expected to fly in the upper sector . . . Nevertheless, there had been a strip: that could hardly be called a *neglect* of the regulations, any more than that described the omission of certain data – such as the time at which an aircraft left the flight level – from the strip. This practice had long been abandoned: it was, as Munjas had said in his own evidence, a 'waste of time'. And lastly, said the Judge, it had been charged that the accused had permitted procedural controllers to operate the radar and that during these operations these controllers had given or had received SSR codes (Squawk), given radar instructions and carried out radar separation . . . all of which was against the rules . . .

174

The Court rejected this charge. Primarily it must be observed that the collision was not caused by misuse of the radar. Even if the elements of this charge were valid, there could be no causal link between them and the collision ...

As to requests for Squawk codes, such requests were common: they might be made by neighbouring control centres, by military controllers wanting to differentiate civil aircraft from their own, or used to identify an aircraft which had left its allotted track. Even technicians could request a Squawk for the sake of an equipment check. Why should a procedural controller not make that request? It presented no danger: on the contrary – and the experts agreed – this was a contribution to safety. Planes sometimes took short cuts – only the radar would show this and enable the controller to warn the pilot ... There had been no evidence that a single procedural controller had given radar instructions or had used radar separation on the day of the disaster. There had been two such instances in 1975 and on both occasions Munjas had informed the Administration and had initiated proceedings against the controllers concerned.

Taking this and all other things into account Delic and Munjas could not possibly be charged with conscious negligence and both men were now pronounced not guilty as charged. They would be freed in accordance with Article 321/3, which required proof that the accused had committed the crimes with which they were charged ...

It was the opinion of the Court that there had been no such proof.

Chapter Thirty-four

Dajcic . . .

Julije Dajcic, too, was to emerge blameless from the charge that he had not conscientiously carried out his supervisory duties and that this neglect was linked to the collision.

His own evidence had been given firmly and with a consistency which drew no challenge from the Court. It *did* serve, however, to heighten the embarrassment of the next accused, Mladen Hochberger . . .

Judge Zmajevic concurred with the claim by the defence that Dajcic was obliged to spend some time at his desk. This was the obvious place from which he would carry out his essential administrative tasks – the organisation of the shift, and co-ordination with other air traffic control centres and services, etc. . . .

It was necessary for him to walk through the control room occasionally to check on the sector operations: but the fact that he was at his desk did not mean that he had consciously neglected any of his supervisory duties.

There were only three further issues, said the Judge: whether Dajcic had given permission for Hochberger to leave the room, and if so, whether this was related to the disaster . . . and whether he should have helped Tasic after noting his situation as the only person on the upper sector console.

On the first matter, the Court had not determined that *anyone* had approached Dajcic between 11.00 and 11.15, or that anyone had asked for permission to leave the control room. There was, therefore, no proof that Hochberger had done so: and Hochberger's own versions of this episode were remarkable for their omissions. He had not mentioned asking for permission to leave during the first hearing: only that he had told Dajcic that Tepes was absent and therefore, that Dajcic knew that only one person was left on the upper sector.

Hochberger had next claimed that after telling Dajcic about Tepes, he went to look for him. Again, Hochberger had not mentioned asking for, or receiving permission to leave. . . .

He had then said that Dajcic *knew* that he, Hochberger, had gone to look for Tepes, although he also qualified this by suggesting that Dajcic might not have understood him. He had simply shrugged, said Hochberger. . . . and he wasn't sure, either, whether Dajcic had actually *seen* him leave the room.

Hochberger had failed to mention these things at the trial: he could not

176

specify Dajcic's position at the time he had approached him, and for the first time, had claimed that he had shown Dajcic a watch, presumably to emphasise Tepes's non-arrival . . .

Taxed with these differences Hochberger had repeated that Dajcic had possibly not understood him, following this with the remark that Dajcic was in any case a strict disciplinarian, unlikely to have given his permission for anyone to leave the control room.

The implication that Dajcic was therefore incapable of comprehending such a request was tersely dismissed . . .

The inconsistency of these statements, said the Judge, gave no grounds for believing that Hochberger had asked for permission to leave or that Dajcic had given such permission. Female members of the control room staff had offered testimony on these points, it was true: but those who had been direct witnesses were quite inconsistent about what they alleged they had seen or heard, while the evidence of those who merely repeated what they had been told by others, was worthless . . .

Dajcic, said the Judge, therefore bore no responsibility for Hochberger's departure, and if Hochberger's absence did not contribute to the alleged overburdening of Tasic then there was no link between Dajcic's responsibility and the disaster.

As to the last of the issues, whether Dajcic should have helped Tasic, it was necessary to repeat that Tasic had *not* been overburdened by his work . . . and even if he *had* been, Dajcic could not have determined that fact from his analysis of the flight schedule. He had not been ordered to make such an analysis and it was not among his duties as chief of shift.

Dajcic had explained to the Court that it would not, in any event, be possible to check on the controller's work load from the number of strips on the console since they might not all be active strips: that it would be difficult to intrude, even to offer help, into the controller's visualisation of the traffic pattern, and that, in the final analysis, it lay with the controller to ask for help if he required it. The experts, Kuscer and Jaksevac agreed . . .

It was not possible to prove any of the charges against Julije Dajcic and in accordance with Article 321/3, the Court would set this man free.

Chapter Thirty-five

Hochberger . . .

Whatever embarrassment had so far been his lot, Hochberger could at least temper that humiliation with a substantial grain of comfort. Judge Zmajevic had already conceded that his absence had not been a factor: what must follow must surely be less, much less, than Hochberger had had cause to fear.

Judge Zmajevic, indeed, made only the most brief comment on Hochberger's decision to leave the console: the controller had not acted as he should have done, said the Judge, nor had he been given permission to leave. These were violations of discipline and would doubtless be investigated by the appropriate authorities. The Court was interested only in the relationship of Hochberger's and Tepes's absence to Tasic's dual role as controller and assistant and to any subsequent neglect by Tasic of his duties as a controller . . . no such link, however, had been found to exist.

The Judge amplified this with a detailed recapitulation of Tasic's activities between 11.05 and 11.10 by which time Tepes had taken up his position, and, again, carefully enumerated the Court's reasons for deciding that Tasic had not been overburdened. He had not done anything but his own job until 11.07 and 40 seconds: that included his previous radio communications with aircraft and the co-ordination with the middle sector. He had made no telephone calls at that time or VHF calls and he had not refused the co-ordination. More – he had indulged in that dialogue now notorious for its touch of humour . . .

Except for that phone call Tasic had performed none of his assistant's other duties: and at the time that he was giving those 'non-precise' instructions to the crew of JP550 instead, stressed the Judge, of taking the correct action necessary to avoid the collision, Tepes was already in his place, and had begun his own work.

There could, therefore, be no link between the 'absence' of Tepes and Tasic's final mistakes . . .

Tasic had corrected his mistakes which showed that he had had enough time to do that: and they had been made either because he had not been concentrating on his work or because he had been indisposed . . . And finally, Hochberger had turned over his duty to Tepes at the doorway of the control room instead of doing that, as required by the regulations, at the work position. It was irregular, yes: but it was not related to the disaster, since the purpose of the handover of duty was merely to establish the facts of the traffic situation.

That had been done. Further, on reaching his work position Tepes had checked the situation by asking Pelin whether Vienna had been given the transfer for JP550, although that transfer had already been recorded. He had had nothing further to do for the next three to five minutes, that was until 11.13 and 30 seconds, and therefore had sufficient time to familiarise himself with his own work: the rest of the traffic situation was not his concern. He could neither give clearances in his capacity as assistant controller, make decisions, nor offer suggestions or information to aircraft. All that was the responsibility of the controller . . .

The charges against Mladen Hochberger were not proved. In accordance with Article 321/3 the Court would set him free.

Chapter Thirty-six

Tepes . . .

The Court must believe Nenad Tepes, said the Judge, since there was nothing to disprove his evidence . . .

He had been scheduled to act as controller on the upper sector but was late in arriving at his post because it had been necessary for him to go to the lavatory. He had heard Hochberger calling him whilst he was there and on his way out of the cloakroom had met Milan Dolezal, the instructor and exchanged a few words with him.

Hochberger had encountered him, says Tepes, at the door of the control room and had told him that everything was in order in the sector . . . the Court had noted some doubt about the time at which Tepes had actually taken his seat at the console, but from the available evidence had concluded that this was at 11.10 a.m. His activities from that time had been described in the findings concerning Hochberger . . .

It is necessary to make clear, said the Judge, that Tepes had quite rightly expected that despite his lateness, *someone* – either the controller or his assistant – would have covered his position until his arrival. Knowing that it was forbidden to leave, he could not have expected that Hochberger would have abandoned the console without permission. The lateness of a substitute would not have justified such a departure . . .

As to the effect of this late arrival on Tasic, now occupying Hochberger's place as controller, it had been claimed that Tasic had made four telephone calls which should properly have been made by his assistant: but this was not true, for Tasic had made only two calls. He had made the first one at 11.5 and ten seconds to inform Belgrade that Lufthansa 360 and Olympic 182 had nine minutes between them. His second call had been made at 11.08 and eight seconds when he had again spoken to Belgrade, and *that* call, in fact, should have been made by his assistant.

It would not do, said the Judge, for Tasic to claim that he had *had* to accept the co-ordination with the middle sector because of the absence of Tepes.

Tasic had taken Hochberger's seat and from that moment he had assumed the responsibilities of the controller: and co-ordination and clearance were included in these responsibilities. These things were not the business of the assistants who had no authority in these matters . . .

That had already been stated: and so had the fact that Tasic could have

180

refused the co-ordination. He did not *have* to accept one more problem ... he could even, had he felt sufficiently hard pressed, have asked another sector to take on some of his traffic.

The prosecution had charged that, after his arrival, Tepes had neglected to familiarise himself with the situation in the sector. That accusation too, had been shown to be groundless. It was obvious, indeed, that Tepes had taken hold of matters quite competently, although he could not have known about Tasic's clearance for JP550 before 11.14 when he saw the slip for that aircraft for the first time: not that he could have raised any query then, since he was busy, telephoning Belgrade to clear Tasic's traffic, Olympic 172 and Beatours 932.

This call, said the Judge, had been begun at 11.13 and 30 seconds, only a few seconds after Olympic had signed off, and while Beatours 932 was still in contact with Tasic. Tepes was, therefore, fully aware of the situation in the sector and of Tasic's traffic insofar as it concerned himself: the fact that he had not wasted any time in beginning that telephone call demonstrated this.

Had the Judge so wished, he might have enlivened this account of Tepes's efficiency with a genuine touch of colour, for at that point, Tepes, like Tasic, had become involved in a minor comedy. Judge Zmajevic, however, who had so consistently pointed to the significance of that interlude in the case of Tasic, did not feel it necessary to single out Tepes now. Possibly, he may have felt that there were no more useful lessons to be drawn.

But, in fact, Tepes had been frustrated in his attempt to call Belgrade and after his opening sentence 'Sarajevo, estimate, upper ...' had spent the next 22 seconds exchanging fruitless 'hallos' with the Belgrade controller. At 11.13 and 38 seconds, he had flung in the towel and said, 'I'll call again ...': and had done so at 11.13 and 52 seconds but had received no reply and finally, at 11.14 and two seconds, had offered the chief of shift an illuminating review of the situation: Julije, he had called mournfully, opet Beograd zajebava ... Belgrade is f ... again ...

Tepes had at last made a successful contact with Belgrade at 11.14 and 15 seconds and ended that conversation at 11.14 and 47 seconds without knowing that the collision had just occurred. He had not, said the Judge, noticed anything unusual during the last minute, nor could he have usefully intervened. All he could see was that Tasic was hunched over the radar which meant that he was either following the track of an aircraft or vectoring: and according to DOC 4444 Attachment C, Article 2.2 page 29 a controller was not to be disturbed during those tasks.

If Tepes had been guilty of indiscipline that would be investigated by the Administration: but the Court had determined that neither his late arrival at his post nor his part in accepting the hand over of the sector at the door of the control room had contributed to the disaster. Nenad Tepes was therefore set free.

Chapter Thirty-seven

Erjavec, Bojan; Pelin, Gradimir ...

It had been agreed by the experts that the work of the middle sector had been 'faultless'. This was confirmed by Samardic, by Stojkovic, by the Commission and by Kuscer and Jaksevac: the Court, therefore, in subscribing to this, decisively reinforced that onus already laid upon Tasic.

That acceptance permeated the Judge's disposal of the seventh and eighth accused, for the greater part of his references to Erjavec and Pelin contained no hint of criticism of those two men: indeed, that their evidence was now to be given credence ensured merely a further savaging for Tasic.

The middle sector controllers, said the Judge, both saw that Tasic was alone when Pelin approached him for the co-ordination ...

Once more the Judge drew the distinction: being alone did not mean that Tasic was overburdened ...

Nettled, Weston heard this amplified: ... and since the Court had determined that Tasic was not overburdened, Erjavec and Pelin could see no indication of that ...

That odd logic could only be noted for another challenge: meanwhile the Judge continued that course wherein the activities of the seventh and eighth accused served as benchmarks against which were measured Tasic's own offences.

Thus, it was accepted that the co-ordination had been carried out in the manner described by both Erjavec and Pelin. It was accepted that Pelin had carried the flight data strip with him and that he had actually handed it to Tasic: that Tasic had fully understood the transaction and that he had been shown the blip for JP550 on the radar screen, following which, he had given permission for the aircraft to climb. The Court had heard this disputed, said the Judge: but when confronted with Tasic, Pelin's demeanour had carried more conviction: and besides, there were those other significant details to corroborate this, namely, that 'Erjavec had pulled the strip a little way out of the rack' for Pelin: plus the fact that Tasic, who had received the strip at 11.12 and 12 seconds 'had not reacted' when he saw that JP550 was climbing to flight level 350. It could only prove that he was not surprised or anxious since he *had* given the necessary clearance. The irregular nature of that clearance, the Judge reminded the Court, had already been noted: and once again came that recital of violated Chapters and Articles ...

The Judge had progressed more than half-way through his findings on Erjavec and Pelin. His words, so far, however, had been so pointedly devoted to sealing the case against Tasic that the revelation of 'further charges' against the seventh and eighth accused came as an arresting reminder that they, too, were being judged.

Erjavec had operated the radar without the necessary licence and, because of his lack of experience, he had sent Pelin ... who had made an incomplete radar identification. In addition, the transfer of radar control and the use of codes had been carried out in a manner contrary to the requirements of the relevant Articles and Chapters: and finally, Erjavec had not given the upper sector code to the pilot of JP550 but instead, had instructed him to Squawk standby ...

The Judge wasted no words in his rejection of these charges: the hand over, he said, had been procedural and not by radar ...

Pelin had stated that he had been helping Erjavec as radar controller at the time: this was permitted – Pelin, of course, was properly licensed ...

Similarly, it had been determined by the Court that Pelin had shown Tasic the target of JP550 by putting his finger on the blip ... there could, therefore, be no question of an 'incomplete identification' ... As to the instruction to the pilot of JP550 that he should switch his transponder to standby, this was not specifically forbidden: there were, in fact, no rules in either domestic or international regulations which covered the temporary switching off of a transponder during the transit of an aircraft from sector to sector. Erjavec had used the Squawk code for his own sector because it made identification easier. Tasic had not given the aircraft a code because, as stated, he had accepted a procedural and not a radar transfer.

These, and related charges, were consequently invalid ... the Court had determined that the work of these two controllers had been entirely responsible: this was endorsed by the expert witnesses ... and therefore, both men were relieved of any costs and set free in accordance with ZKP, Article 321/3 ...

In this case, the Judge observed, the Court had learned something of the work of the air traffic controller: enough, perhaps, to understand that it was only necessary for one man to break the rules in order to bring catastrophe near.

A single transgression or a single error might be enough: the foundations of disaster required no more ...

Sadly, the aptness of that comment would be marred in the future by debate as to its proper target: but in this case, eight men had been accused and seven, albeit forever scarred by the experience, had emerged into the daylight. Only Tasic now remained to be thrust into a deeper shadow.

Chapter Thirty-eight

Weston had caught the essence of this trial in that reference to its thread of pain: for that suffering, although shared by hundreds, remained a unique torment for each individual.

It would be borne thus by each member of the families of the victims and by their friends: by Weston himself and, up to the moment of their exoneration, by the seven men of the air traffic control service, by *their* families and by their friends. These people could and now did rejoice as in the lifting of a dreadful affliction: but it could not be turned aside for Tasic, for his wife Slavitsa, and for those who were near to them.

For these, the crisis of this adversity was at hand and they would now savour the special distress of seeing the record stamped with an imperishable irony.

It was clear, said the Judge, that in all of this there had been no element of wilful mischief: nevertheless, the matter of negligence was inescapable and as a controller Tasic must have been aware of the consequences of that negligence: more, he had showed that negligence towards the existing rules from the moment he had made that first basic mistake, namely, when he had given permission for JP550 to climb, without making any conditions.

The Court, therefore, had found this man guilty: but of what did that guilt consist?

Guilt was not something real and objective. It was an abstract thing, a value judgement escaping rational analysis. The step from guilt to punishment was a step from the metaphysical to the empirical . . .

It was, however, the central notion of criminal law and the legal expression of the personality of the individual tried: it applied to Tasic, the Judge pointed out, despite the claim by a representative of an injured party - Weston grimaced as Modrusan explained this - that practical reasons spoke against the idea of responsibility for negligence and that the basis of guilt should be intent and damaging action. Judge Zmajevic rejected the concept as unrealistic. Of course, reprisal was not always an adequate response to negligence: of course, such a punishment could be pointless: but what was being put forward by that claim was that negligence should be removed from the criminal law entirely and *that* would leave a great deal of dangerous behaviour unpunished.

Tasic would be punished, said the Judge. It was necessary, in coming to such a decision, to remember that as between that pointless reprisal he had mentioned

184

and the complete absence of punishment for negligence which had been suggested, the law must take the middle course: and it must do that because it was not possible or desirable to look upon our modern technology either as a guarantee of absolute safety or as a licence to live dangerously ...

The Court had accepted that there was no element of wilfully criminal behaviour as proposed by the prosecution: yet Tasic's negligence had resulted in all the hazards foreseen in the Articles of Indictment, detracted from the efficiency and standards of the Zagreb Flight Control Centre which, until that time, had satisfactorily performed its duty, and finally, had brought about the catastrophe.

The point had been made that as a responsible controller, Tasic must have been aware of the possible consequences of any deviation from the rules and that such deviation must involve punishment ... and the idea of punishment, Judge Zmajevic impressed on the Court, was a device justified by experience, for the strengthening of social responsibility. Yet no one should be in doubt, he warned, as to the basis of such punishment as was warranted in this case. A catastrophe of this kind became a world incident: it was the public who insisted on identifying those responsible for it and very often, the public even believed that it was its duty to create such figures. There were occasions, indeed, when social pressures were blamed for the problems or the downfall of an individual or when the public apparently felt that it was necessary to harbour some kind of guilt complex in respect of those who committed criminal acts. ...

Was that not evident here, in the claim that it was the flight control system itself which was at fault, despite the fact that this self-same system had safely guided 760213 aircraft through its airspace between 1971 and 1976?

One could not support such attitudes. Yet public opinion had to be taken into account: not in terms of its emotional demands but in the interests of that confidence which the public was entitled to place in the laws which protected it.

It is my task now, said the Judge, to sentence the accused, Gradimir Tasic.

His punishment is decided, not only by the aggravating circumstances, but also by the extenuating factors, by the personality of the accused, the circumstances and seriousness of his act and by his attitude.

The Court has noted and considered among the extenuating factors the delay on the part of the crew of JP550 who had been given the upper sector frequency at 11.12 and 12 seconds but had not made contact with that sector until 11.14 and four seconds, that is, after one minute and 52 seconds.

It is laid down in Annex 11 of the rules of the International Civil Aviation Organisation, Chapter 3.5.3, that such calls must be made as soon as possible: yet the captain of JP550, the person responsible for the flight, switched his radio to an unknown frequency four times, at 11.09 and 53 seconds, at 11.11 and 46 seconds, at 11.14 and 20 seconds and at 11.14 and 24 seconds. And on the last two occasions he should have been switching from the middle to the upper sector.

185

We know from the report by Inex-Adria that the captain spoke to the operative centre between Split and Kostajnica, although the exact time of communication was not recorded: it is not, however, of importance. It is only relevant that this aircraft, JP550, did not make contact with the upper sector controller at the correct time.

As for the other aircraft, said the Judge, the Trident – the crew were also obliged by ICAO Annex 11, Chapter 3.5.5, to listen out on the operating frequency ... They must, therefore, have heard the message from JP550 at 11.14 and ten seconds, that the aircraft was passing flight level 325 estimating Zagreb at One Four ... That message was in English and although the representative of British Airways was right to claim that the conversation which followed was in Serbo-Croat and so without meaning for the British crew, nevertheless they had heard and understood that first message and should have kept a careful look-out ...

Unjust, thought Weston. He noted, too, the blurring of fact and responsibility in the next words: It is recognised, went on the Judge, that the closing speeds of the aircraft made it difficult for the pilots to see the machines in time: but it is true, also, that the Trident was leaving a condensation trail behind it and that the collision took place in conditions of the clearest visibility.

Gradimir Tasic has never been convicted previously. He is married and is the father of two children.

Again on familiar territory, the Judge allowed his tone to signal the ending of that long wait: there had, he said, been no family difficulties ... but there had been certain subjective problems, notably connected with Tasic's housing, which had existed during the period of his employment at Zagreb. He had had no apartment of his own and, of course, had been given that accommodation in the disused radio hut ... Because of this, his wife had returned to Belgrade and Tasic had pressed the authorities for his own transfer to the Belgrade air traffic control staff. This had been approved, shortly before the accident: but no doubt, that news had contributed to Tasic's unconscious indisposition for work on that day, since controllers who had been transferred, because of the needs of the service, from Belgrade to Zagreb were often under an extra pressure as their thoughts wandered between their families in Belgrade and their work at Zagreb ...

Weston greeted this intelligence sourly. That possibility had not seemed to have unduly troubled anyone up to this point. Only Tasic ... who was, said Judge Zmajevic, in this unfortunate position: and Tasic's work had been made even more difficult by the fact that, although he held a radar rating, the radar at the Centre was on test and unreliable. It had also been obvious during the trial that Tasic lacked certain technical knowledge concerning aircraft performance and this constituted yet another of the pressures on him.

Very well, said the Judge. The Court had considered all the elements of this case ...

It is my decision that Gradimir Tasic should be sentenced to seven years' rigorous imprisonment. He has been in custody since 10 September 1976 and the period from that date is reckoned in the length of his sentence –

The Judge quelled the sudden stir of movement in the courtroom.

– the accused, he ended, lives in poor material circumstances and has no prospects. According to ZKP, Article 91/4, the Court has therefore freed him from paying the expenses of this trial.

Chapter Thirty-nine

The jailing of a hitherto unknown air traffic controller in Yugoslavia would not normally have created more than a ripple of interest: Tasic, convicted, might well have disappeared into that limbo which so readily swallows others.

That he was not permitted to do so – that at length, but only at length, and only after he had grown all too familiar with the bitterness of imprisonment and his wife equally familiar with the fact of their sundered lives – was due to the solid support of the outraged membership of the International Federation of Air Traffic Controllers' Associations and the persistence and determination of that body to overturn what they saw as a grossly misplaced verdict. Those efforts were to be personified by Richard Weston, now formally retained by IFATCA to mount an appeal, and by his two contacts in that organisation, the President, Jean-Daniel Monin and the General Secretary, Tom Harrison.

For all its sincerity, for all its justification, that indignation could not hope to touch the lay public: in such a context the outcry from the professional community was a muted one and was certainly far from being the 'storm' of protest headlined in one British newspaper. That word did, however, accurately describe the vigorous reaction of those concerned: the protests were certainly no less valid for their limited readership. They would also strike chords to ensure that Tasic's story, so fleetingly made known, would not afterwards be lost in either the impersonal text of the Accident Report or the files of the legal record.

Weston himself made the first comment for the assembled journalists: 'feelings', he told them, 'would run high among controllers ... the prison sentence pronounced here adds to the already great pressure on flight controllers throughout the world. If there is to be the threat of legal proceedings hanging over them it will have a very bad psychological effect.' At this point, too, Weston indicated his own resolution: 'I hope that international action in the form of an appeal by flight controllers will overturn the verdict on Tasic and secure his release. This sentence sets a dangerous precedent and is entirely counter-productive: it would have been much better and more just to have found out what is wrong with the system, rather than to blame everything on an individual.'

These sentiments were put more forcefully that evening by Gordon Rowland-Rouse of the British Guild of Air Traffic Control Officers who described the sentence of seven years' rigorous imprisonment as 'shattering:

188

Tasic has been made the fall guy. He was hopelessly overworked by our standards of safety.'

A more temperate letter appeared in *The Times* 20 May 1977, over the signature of Mr John F. Blunden:

> Most of us will have learned with horror and astonishment that the air traffic controller found guilty of negligence and therefore held to be responsible for the terrible collision over Yugoslavia last year has been gaoled for seven years.
>
> Anyone concerned with the maintenance of technical standards where life is at risk will have met the problem of how to deal with a human failure. Deliberate sabotage is, of course, an entirely different matter and is an obvious criminal charge, but the man who has a momentary failure while doing his duty, caused in all probability by stress or other reasons beyond his control cannot have earned imprisonment.
>
> The (armed) services in such instances generally resort to reducing the guilty person's status by loss of rank and thus, privileges. It has the effect of stressing to others the severe responsibilities they hold. Judgement in such cases is often difficult and frequently unfair, but to deprive a man of his freedom . . .
>
> As a leading aviation country the United Kingdom should appeal immediately to the Yugoslav authorities for clemency. The fact that so many British subjects were the victims would add weight to this appeal.
>
> I cannot believe that any bereaved relative of those tragic victims would wish it otherwise.

It was still not a storm: but while there was yet some public interest Weston moved to head off one imputation which could only breed a justified resentment in Yugoslavia. In a letter to *The Times* 28 May, he wrote:

> . . . having been present during most of the trial I wish to offer the following observations for the attention of a rightly concerned public:
>
> 1. Although conducted in a manner notably different from our own, the trial itself was fair by any standards. The rights to hear and be heard, to examine and to cross-examine, were both granted and exercised to the full and – as would not be the case in this country in a criminal proceeding – even I, representing a party aggrieved by the alleged offence, was granted and exercised these privileges.
>
> 2. In so far as one could ascertain, all the available evidence was painstakingly collected and presented to the court by those concerned with the case and once again in so far as one can pronounce on these things it was received and considered by the court impartially.
>
> 3. This was done in public with full opportunity for both domestic and foreign press coverage (except for certain sessions in camera when military

189

evidence concerning air traffic control was heard – and even then I was permitted to remain in court).

4. While one may not agree with the verdict, particularly in relation to Gradimir Tasic – and I do not for many reasons – I believe it to have been reached by a due judicial process.

5. Any judge or, for that matter, judiciary must operate within the parameters of that system which he or it serves: to that extent, I believe the acquittal of seven of the eight defendants was a morally courageous decision by Judge Zmajevic . . . (but) moves which I do not feel at liberty to discuss are currently in hand to correct what I and many others feel is a mistaken determination in the case of Tasic.

Among the vanguard of those others was the United Kingdom Guild of Air Traffic Controllers. Len Vass, acting as Public Relations Officer, set out their case in a letter to the aviation journal *Flight International*:

. . . 26th May, 1977

The verdict and the sentence imposed seem harsh . . . baffling to most British observers in view of the commendably efficient and courteous manner in which the Yugoslav authorities conducted the investigation.

The collision had an electrifying effect on the ATC profession all over the world and controllers have given much attention to the events which were either known or suspected to have contributed to it. Much of this interest is based on the feeling that there, but for the grace of God, they themselves might one day go.

As a breed, controllers are notoriously sensitive to (any) operational misfortune that may befall their colleagues and even more so when system deficiencies are identified and coupled to high workload factors. Certainly, any national authority must retain the right, in the interests of flight safety, fully to investigate ATC malfunctions. Where glaring mistakes are made, steps must be taken to prevent a recurrence when errors can be attributed in full or part to system error . . . However: when there appear, as in the Zagreb case, to be a number of human errors contributing to an ATC incident in a profession which tends to get more than its fair share of stress-induced and multi-dimensional difficulties, then prison is hardly (an appropriate) penalty for default . . .

Vass ended by stressing the very crux of the matter:

If punitive action is deemed to be necessary for such 'crimes' it surely would be more just for any country to make it a matter for internal discipline not geared to the criminal mode: and even then, only where flagrant disregard of procedures for no valid reason have been proven. The Guild of Air Traffic Controllers hopes that Yugoslavia may be encouraged to reconsider both the verdict and the sentence on Tasic, who, it feels, has been the victim of circumstance coupled with air traffic control system

190

deficiency ... This man is not a criminal.

... If clemency and a greater depth of appreciation and human understanding are applied to the dreadful situation in which he found himself on that occasion then Yugoslavia's international aviation standing can only be enhanced by that.

Conversely, if confirmation of the sentence is upheld ... she will effectively have raised a veritable Sword of Damocles over every air traffic controller, eroded confidence within her own ATC service and set a precedent for those Administrations whose standards of perception and justice may not be as exact as those which we, in Europe, normally associate with Yugoslavia.

Vass's letter was clearly a genuine *cri de coeur*. It nevertheless triggered, or perhaps helped to perpetuate, an unhappy and sterile rift between two bodies with a complete identity of interest in the matter.

On Weston's initiative the appeal was also to be supported by a formal Petition to Marshal Tito, President of Yugoslavia, requesting a reversal of the verdict and the release of Gradimir Tasic.

Signatories to the Petition included Jean-Daniel Monin and Tom Harrison for the fraternity of controllers embraced by IFATCA and John Leyden, President of the Air Traffic Controllers of America (PATCO) on behalf of his own 17 000 members. The hope of a concerted protest by the aviation community could not, however, surmount the pitiable state of the relations which existed at that time between IFATCA and the pilots' counterpart, the International Federation of Air Line Pilots Associations (IFALPA). Tom Harrison commented on that foundered aspiration in the circular of August 1977 ...

We are sad, though not very surprised, that some world organisations and IFALPA in particular with whom we work so closely in other fields, found themselves unable to sign and seal the document after some negotiations had taken place ... Most regrettable ... but at least we know where we stand. Common adversity has struck a chord with PATCO and we are delighted to have them on our side once more ...

The plaint provoked an angry response from Captain Laurie Taylor, Executive Secretary of IFALPA:

20th September 1977 ...

... have to inform you that the first page of your August 1977 circular is, in my opinion, grossly unfair ...

I believe that our representations were made to President Tito earlier than your own and may well prove to be equally valuable to Mr Tasic. The text of the IFATCA draft statement by Mr Weston was too long and too particularised for our taste, and so Captain Pearce wrote a very different letter to President Tito ...

Captain Taylor offered a more detailed criticism:

> 'Swords of Damocles' . . . and similar cliches and purple prose are unlikely, in our opinion . . . to be given serious consideration by a Head of State or by others able to influence events. Perhaps you will see your way clear to inform your Member Associations that IFALPA made earlier and separate representations which were intended to effect the release of Mr Tasic and to avoid a similar fate befalling any other controller – or pilot – in Yugoslavia . . .

Nevertheless, two major moves had already been set in train: within days of the affirmation by Len Vass, the International Federation of Air Traffic Controllers' Associations had declared both its own position and the adoption of Weston as its champion. The IFATCA circular of June 1977 addressed to member associations in 50 countries stated:

> The result of the trial sentencing Gradimir Tasic has shocked the controller world and poses many questions which need answering. There is still an appeal to be heard and readers should be assured that IFATCA is very deeply concerned that one of our members should have been sentenced to seven years rigorous imprisonment.
>
> Pending the appeal the Executive Board has made no public statement although it is true to say that Member Associations have been greatly dismayed to the extent of proposing a world strike at a later date.
>
> No aviation administration should feel complacent about this verdict and those who run their air traffic control services on a shoestring or with minimal or substandard equipment and conditions for air traffic services staff . . . should now be worried lest they find themselves involved in a similar horrifying accident . . .

The announcement ended by introducing Weston to the membership as 'a British Solicitor who was prepared to come to the aid of Gradimir Tasic' and noted, in addition, his election as an Honorary Member of the Yugoslav Air Traffic Controllers' Association.

That tailpiece marked the opening of what had now become a personal crusade: it would entail much struggle, see some success of a kind, and demand the persistence of near obsession. Nothing less could sustain an effort it would be necessary to exert in the face of continuing difficulty for many years.

Chapter Forty

From Richard Weston
London

September.

Dear Ronnie
- A long time ago we met for the first time to talk about Gradimir Tasic. I think now, looking back over these pages, that we've done what we agreed to do on that occasion.

Primarily we wanted people to understand what had really happened to Tasic, what kind of pressures controllers worked under and what it could be like to have to make their kind of decisions.

It won't be possible, I think, for anyone to read this without coming to such an understanding and without recognising just what an enormous and bitter irony was the fact of this man's imprisonment.

Yet while I believe that that feeling was shared by everyone involved in this case - and I include the Yugoslavs - I am not sure that one can speak of injustice. Whatever their own feelings, - and I know that I have said this before - the Judge and the Prosecutor carried out their duties under the law and it is that recourse to the criminal law in such cases that we must challenge if the verdict on Tasic is not to be accepted as a precedent.

This then, is part of that protest. As to 'changing' things, Ronnie, we cannot look too far into the future but hopefully, this book may one day help to tip the balance with responsible aviation people: one day the choice may again involve the honest and public acknowledgement of impotence, or even the revelation of indifference in the face of system deficiency - of sub-standard, ill-matched or even non-existent equipment, of inadequate or ill-trained staff and of working conditions which may be totally unsuitable and even positively hostile to the controllers' needs. .... The confession of any or all of those things is one of the choices and, given any kind of decency, let alone regard for aviation safety, their rectification must follow. The other choice is the one we have seen exercised in the case of Gradimir Tasic.

We will not expect miracles, my friend, but at least we have put the issues fairly and perhaps managed to say something people will want to think about. That other choice is still available in too many parts of the world: but again, hopefully, it may not now be made so readily and it cannot now delude either those who make the decision or those for whom it is intended.

We seem to have left some loose ends: I suppose that phrase is particularly appropriate since even now, in 1982, the repercussions of the Zagreb disaster are still echoing through the Courts. . . .[1]

But let us go back to '77.

By this time I had become very deeply committed. I had been dealing with Ruth's estate, of course, and with all the other matters arising from her death – the file will give you some idea of the volume of correspondence and the quite astonishing number of things which it was necessary to do. It will also reveal what was involved in mounting the Appeal against the verdict and in organising support for the Petition to Marshal Tito.

We laboured through all the formalities of the Appeal – I think that is a fair description – but by the early part of 1978 the whole thing seemed to have become bogged down: nothing seemed to be coming back to us except the fact that Tasic was serving his sentence. By that time, too, I'd made a number of very frustrating visits to Yugoslavia, seeing various people in the Justice Ministry and being told consistently that there was nothing they could do to interfere with the process of the courts. That was clearly true but it didn't make it any easier to accept.

But in 1978 the International Federation of Air Traffic Controllers' Associations was to hold its annual meeting in Copenhagen. The Appeal had *still* not been heard and as a result the controllers felt it necessary to consider some drastic action and that would certainly be on the Agenda.

I had been invited to address the conference and in another bid to get something done before then I went back to Yugoslavia to see the people at the Justice Ministry. I told them quite bluntly that the International Federation was not prepared to sit idly by and see one of its controller members put in gaol without lifting a finger to assist him. Things had been dragging on for almost a year since the date of the conviction and sentence – and in fact, of course, Tasic had been held since 10th September, 1976 and here we were in April 1978 –

Look, I said, this man has been in gaol now for eighteen months: what's holding up the Appeal?

I told them about the feeling among the controllers and that Copenhagen was going to be something of a watershed: I didn't want to make it seem like blackmail but on the other hand it was more than time something happened. The thing had gone far enough and this was an opportunity to say so.

Clearly, we got the message over. I went away in a better frame of mind and jumped the gun somewhat by sending a telegram to Drasko for the attention of Dr Dusan Cotic, Deputy Minister of Justice. In that I asked for the immediate release of Tasic on parole. If there was going to be any possibility of that outcome I wanted the news to be given out at the Conference. I sent that telegram on 18th

---

[1] In 1981 relatives of the victims began an action in the California Courts against McDonnell Douglas. It was alleged that the design of the DC9 cockpit window pillars was such as to impair the pilot's capacity to look out for conflicting aircraft. At the time of going to press the case was unresolved.

April, reminding Dr Cotic that the conference was to begin on Monday, 24th of the month: but despite the indications I had been given that things were moving in the right direction at last there was a tremendous sense of anxiety for the next few days . . .

But, in the course of the conference – I believe it was the 25th – I received a cabled reply:

> The Supreme Court of the Socialist Republic of Croatia has made a decision on Tasic's case. A verdict will be given at the end of the next week.
>
> I assure you that there is no reason for any doubt about justice (in) Yugoslavia. Every circumstance pointed out by IFATCA will be taken into consideration. Greetings. Cotic.

It was still too early to rejoice. We all did that, very briefly, on the 29th April when the decision of the Supreme Court was officially announced. Tasic's sentence had been halved and he would now serve three and a half years.

But as far as IFATCA and I myself were concerned that was only one milestone. It was welcome enough, of course, and what was particularly welcome about it was the absolutely unambiguous statement in the Appeal verdict that '. . . mitigating circumstances included imprecise rules and selection and training methods for flight controllers, overloading of the Zagreb regional flight control and' . . . 'a number of other circumstances for which the greatest part of the responsibility lies with the Yugoslav Civil Aviation Authority'.

The Appeal judges repeated, however, that the accident '. . . would not have happened had Tasic acted in accordance with the prescribed rules and his professional training' but for the moment anyway we were so happy for him and for the Judges' recognition of our case that we let that one pass. But only for the moment.

What we *did* do, though, was to get down to the Petition in earnest. Far from serving three and a half years we were all determined that Tasic would not spend one more day in prison than was necessary to get him out: if we were going to succeed in that then we needed to keep up the pressure.

It was, after all, a matter of nine months since Jean-Daniel Monin and I had delivered the Petition: to be precise we did that on the occasion of a meeting with the Minister of Justice of Yugoslavia, Mr Ivan Franko on the 8th September 1977 at 12 o'clock. We could not possibly have hoped for a dialogue at a more influential level, since the Minister was supported by Mr Vrazalic, Assistant to the Federal Secretary, Mr Josifovic, Chief of the Cabinet to the Federal Secretary, and Mr Gajic, Director of the Federal Organisation for Civil Aviation: it was I think a most splendid and generous acknowledgement on the part of the Yugoslav Government, of the gravity of the whole thing . . .

We spent the best part of an hour with them and put the IFATCA case. I should doubtless say 'yet again', but they were courteous and sympathetic, nevertheless, and told us frankly what the situation was at that time – that the matter was then in the hands of the Supreme Court who were not only considering the appeal by Tasic, but in addition, that of the Public Prosecutor against the acquittal of the seven other controllers. They hoped to have a decision, they said, before the end of the year: meanwhile the Minister would study the Petition and IFATCA would be informed of his consideration.

Which, for the time being anyway, had to satisfy us. As a matter of interest by the way, Ronnie, we emphasised the need to remove the Sword of Damocles which now hung over every controller: I fear now that you will suspect me of being mischievous in this but I assure you it was not so. And happily, the Minister and his colleagues were not dismissive of the analogy.

The verdict on the Appeal showed that we had achieved something: but we were agreed, as I have mentioned, that it was by no means sufficient and by June, '78 we were probing for the best possible course of action. There had not yet been any response to the Petition: indeed it seemed to have disappeared into the Justice Ministry despite the fact that IFATCA had been bombarding the Marshal – or at any rate, his official address – with a stream of telegrams pleading for his attention.

In my own perplexity, therefore, I wrote to people in Yugoslavia, offering to go back again if it would help to expedite things. I wrote to Toma Fila, the Advocat of Gradimir Tasic (Fila, incidentally, had defended Tasic without fee and throughout had displayed a genuine concern for the welfare of Tasic's family) ... to Zeljka Modrusan, Drasko's sister, who had been of tremendous help to me there and to Deputy Minister of Justice Judge Predrag Backovic. I offered my sincere congratulations to Fila on his part in achieving the reduction of Tasic's sentence and asked him what other legal avenues were open to us. There were, I thought, at least two: Fila could ask the President to consider the Petition and to grant a pardon to Tasic. That word 'pardon' was used with reluctance since it implied the commission of an offence: but the main thing was to obtain Tasic's freedom.

The second course was to ask the Attorney General of the Republic of Yugoslavia to make a further appeal to the Supreme Court for a reversal of the decision by the Court of Appeal. That may look odd: but we wanted more than the reduction of the sentence: we wanted it to be quashed. I argued that such an appeal might be made on the basis that that decision was, in fact, bad in law: if the Attorney General would go along with that, then I wanted to support the action.

To Judge Backovic I wrote that the Court of Appeal had made its decision to reduce the sentence from 7 years to $3\frac{1}{2}$ years solely on the facts put forward by the parties at the original hearing: at this stage, I pointed out, the matters contained in the Petition had not been considered. I wanted to ask the Judge if that consideration could now be given by the Head of State, Marshal Tito. If the judicial process had indeed come to an end, any action by the Marshal or by the Minister acting on his behalf would not be counted as an interference.

And, of course, I wrote to Zeljka Modrusan enclosing the letter to the Judge and asked her to make sure that it went to him because I was having trouble tracking down his official address: I had written to Dr Cotic previously and had had that letter returned to me.

There has been little occasion to smile at anything in this book, Ronnie, but there was something quite extraordinary in the way the impasse was broken. Or at least, we may choose to give that weight to what actually happened although I am sure that there will be a more formal, and perhaps more credible, version of the episode in the official record.

But in March, President Tito's aircraft landed at Shannon airport for refuelling. The Marshal was en route to the United States for a meeting with President Carter and of course the arrival at Shannon was very much a VIP matter.

Writing to Mr J. Kearns of Shannon Air Traffic Services afterwards, Tom Harrison expressed his delight that Kearns had managed to get the agreement of the Irish Department of Foreign Affairs to enable him to present a copy of the Petition to Marshal Tito ... 'as he passed through the airport yesterday morning'. It was indeed, wrote Harrison, ... 'a most opportunist thought and effort which I hope will make the grade...' Coincidently, said Harrison, he had heard from Weston that the Public Prosecutor had dropped his Appeal against the acquittal of the other defendants ... 'I told Weston about your intended effort and he was equally pleased and said could we not place a copy of the Croatian version of the Petition in the Marshal's hand, when he returned?

'I discussed this with Pat O'Doherty and he suggested I send you a copy direct in the event the Marshal will be returning via Shannon. I leave it in your good hands ...'

So much for the record. What I am told actually happened is that the Irish controllers, appraised of the Marshal's presence, simply said in effect that until he personally accepted a copy of the Petition his aircraft would not get a take-off clearance ...

It is more than possible that this *was* suggested. I am not sure though, that they really thought that Yugoslavia would be content, in that case, to conduct her international affairs from a stationary aircraft at the end of a Shannon runway: but in the event, the matter was not put to the test. The Marshal agreed to accept the Petition from the hands of the Yugoslav Ambassador to Ireland who had arrived at the airport to meet him: but not, I am told, without some indication of the Presidential displeasure. I understand, Ronnie, that this is a euphemism: according to my own informant he was hopping mad.

It is strange, perhaps that we should be ending our work on so light a note but it is not entirely inappropriate. For the most part we have recorded a good deal of sorrow and not a little that was quite terrible: but there is something more.

Gradimir Tasic was released from the Povarevac Prison on 29th November 1978. He had served some two years and three months.

It is true that his release owed so much to the efforts of colleagues and well-wishers in very many countries: but I want to say, too, that it could never have been granted without the intercession of those people in Yugoslavia who became increasingly uncomfortable with the verdict, who realised its irrelevance and who were concerned for their country's international reputation. So we have to add that to everything else.

And Tasic?

Tasic is building another life. We all hope that it will bring him peace of mind because that is his right. It is something for a person to know that people cared and did not forget him from the very first moment of all that trouble.

We began this book by painting the picture of Tasic's face reflected in every radar screen. We have ended by showing that there is a moral decision to be taken by legislators and aviation administrators alike if that image is ever to fade.

When it does we will all feel better about boarding an aeroplane: and that will be when every conscientious controller, in any part of the world, can be certain of one thing ...

What happened to Gradimir Tasic is not going to happen to me.

Richard